Rodents

Edited by Loth S. Mulungu

Published in London, United Kingdom

IntechOpen

Supporting open minds since 2005

Rodents
http://dx.doi.org/10.5772/intechopen.83079
Edited by Loth S. Mulungu

Contributors
Samson Arockianathan, Tatsuo Yabe, Roberto Carlos Vera, Israel Muñoz, Vilena Kašuba, Vedran Micek, Alica Pizent, Blanka Tariba Lovaković, Davor Želježić, Nevenka Kopjar, Mirta Milić

First published in London, United Kingdom, 2021 by IntechOpen
IntechOpen is the global imprint of INTECHOPEN LIMITED, registered in England and Wales, registration number: 11086078, 5 Princes Gate Court, London, SW7 2QJ, United Kingdom
Printed in Croatia

British Library Cataloguing-in-Publication Data
A catalogue record for this book is available from the British Library

Additional hard and PDF copies can be obtained from orders@intechopen.com

Rodents
Edited by Loth S. Mulungu
p. cm.
Print ISBN 978-1-83962-826-9
Online ISBN 978-1-83962-827-6
eBook (PDF) ISBN 978-1-83962-828-3

Meet the editor

Dr. Mulungu's research interest is in pest management and ecology for both invertebrates and vertebrate pests. Generally, pests cause serious agricultural and health problems for people. Studying the ecology and biology of pests can lead to significant improvements in the way we and farmers manage pests. Ecologically based rodent management (EBRM) and Integrated Pest Management (IPM) are increasingly seen as more sustainable, both economically and environmentally, than the traditional use of synthetic poisons. One of the big problems in developing better pest management strategies is to understand their true impact on people's livelihoods. Although many farmers will understand that many pest species have problems and damage their field crops, stored food, and personal possessions, awareness among farmers about the level and scope of the damage is often underestimated. As an expert, researcher, and trainer academician at the Sokoine University of Agriculture, Morogoro, Tanzania, Dr. Mulungu has worked extensively investigating various pest species' ecology, conservation, and management. He is the author of more than ninety papers in refereed scientific journals and books, and has made many contributions to symposia, conference proceedings, and international scientific meetings.

Contents

Preface

This edited volume is a collection of reviewed and relevant research chapters concerning developments within the field of rodent study. It includes scholarly contributions by various authors and experts that examine rodent behaviour, control, and management.

The book includes the following chapters:

- "Prologue: Commensal Rodent Problems Across the Globe" by Dr. Tatsuo Yabe

- "The Influence of Electromagnetic Fields on the Behavior of Mice" by Drs. Roberto Carlos Vera and Israel Muñoz

- "Effect of Diet and Water Availability on *Rattus norvegicus* (Rodentia: Muridae) Distribution" by Dr. Tatsuo Yabe

- "Nesting Behavior of Indian Giant Squirrel (*Ratufa indica* Erxleben, 1777) in Mudumalai Tiger Reserve, Western Ghats, Southern India" by Dr. Samson Arockianathan

- "DNA Damage and Glutathione Peroxidase Activity in Liver and Kidney Cells in Wistar Rats Exposed to Terbuthylazine (TERB) for 28 Consecutive Days" by Drs. Vilena Kašuba, Mirta Milić, Nevenka Kopjar, Davor Želježić, Blanka Tariba Lovaković, Alica Pizent and Vedran Micek

The target audience comprises scholars and specialists in the field.

IntechOpen

Chapter 1

Prologue: Commensal Rodent Problems Across the Globe

Tatsuo Yabe

1. Introduction

Black rats *Rattus rattus* have become a rare species in northern and central Europe [1–5]. Most of the USA and Canada are also free of the rats [6]. In Australia, house mice *Mus musculus* pose particular problems in high-rise buildings and skyscrapers [7]. In Japan, house mice and Norway rats *R. norvegicus* are rare in buildings in the centers of big cities. Whereas black rat infestations have been experienced even in modern buildings in almost all big cities since the 1970s except those in the northern area [8–11].

However, black rat problems in big buildings tend to be reduced in these years by structural improvement (T. Tanikawa, pers. comm.).

Instead, black rat problems in residential areas in the centers of big cities have become problems since the mid-1990s [12]. Norway rats living outside buildings in busy streets also have become notable problems [13]. Commensal rat problems in islands in Japan have been focused on from the viewpoint of conservation of ecosystems since early in the 2000s [14, 15].

2. Changes in species composition

From the end of World War II to the 1960s, Norway rats appeared to overtake black rats with urbanization, though total number of both species decreased [16, 17] (**Figure 1**). At that time, sanitation was relatively poor, and 1–2 story buildings occupied even the centers of big cities. Catering establishments such as restaurants and drinking houses were generally in such small buildings and were commonly invaded by Norway rats.

However, the situation changed from about the 1970s. In buildings in the central area of Tokyo black rats became dominant, though Norway rats remained in sewers and in parks and gardens in the area. They rarely entered buildings. In an area in which a pest control company (PCO) operated, black rats infested 66% of all types of buildings, with Norway rats or mixed species of commensal rodents infesting the rest [18]. A similar situation was reported widely in Japan, including Sapporo, a city in the northern area [8–11]. Later, however, the black rat problem disappeared in Sapporo. Causes of the disappearance is not known yet, but I suppose one of the causes is the lower temperature of Sapporo [8, 9]. Accordingly, black rats are dominant in buildings in all major cities except Sapporo.

From around the 1970s, the Japanese economy grew and big buildings with three floors or more rapidly increased in commercial districts [9–11] (**Figure 2**); many were connected with the catering industry. At the same time, though PCOs rapidly multiplied, black rats became successful in these areas. From the mid-1990s, these rats scattered to residential areas. Questionnaires showed that 23% of 322 residences in a ward in Tokyo experienced black rat invasions within the past two years [12].

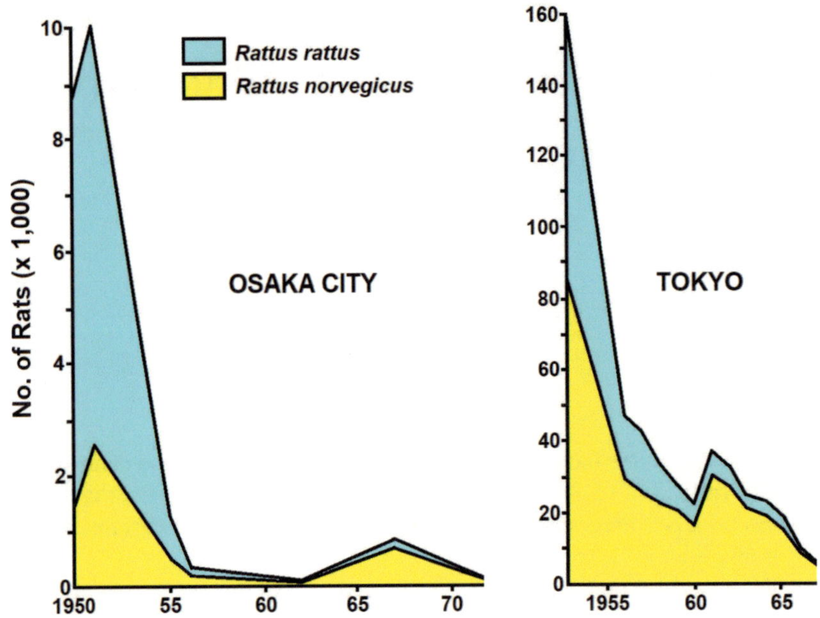

Figure 1.
Changes in species composition of commensal rats collected with snap traps in Osaka city and the Tokyo Metropolis during control campaigns [16, 17].

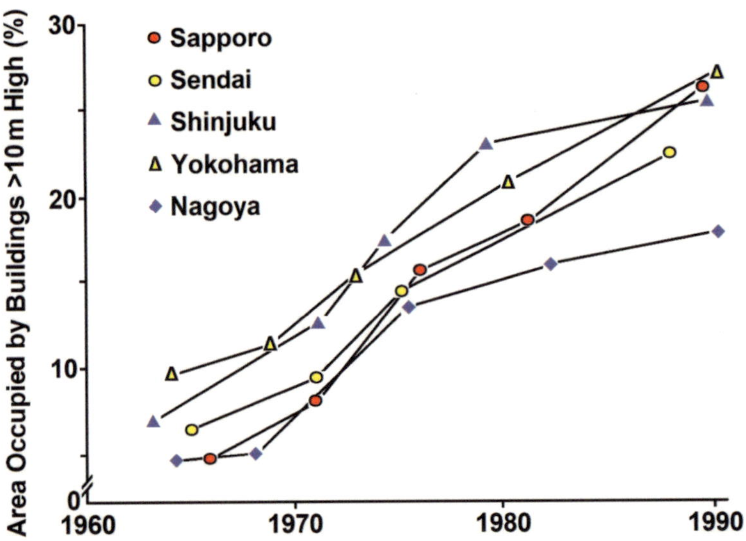

Figure 2.
Yearly changes in percentage of basal area of buildings >10 m high in commercial districts in Sapporo, Sendai, Shinjuku, Yokohama, and Nagoya [9–11].

3. Cause of infestation of black rats

It is the structure of the building that is the cause of heavy infestations by black rats. Big buildings in the centers of big cities usually have several catering establishments and these provide the food for black rats. Such buildings, with their network of hidden pipes and false ceilings, also provide ideal habitats and nesting places for the rats. Warmth from cooking in the catering establishments and from electrical equipment being used in such buildings also probably supply sufficient heat for all-season breeding.

Another cause of heavy infestation is the difficulty experienced in controlling black rats. Most PCOs had long used first generation anticoagulants, and as a result, black rats acquired resistance to the rodenticide [19]; black rats are also extremely cautious of baits. Moreover, catering establishments disapprove of rodenticide operations because of difficult-to-manage rat carcasses that result. PCOs have therefore to use glue boards and of these black rats are also cautious. Today rat proofing techniques are common in PCOs, but long-term rat-proofing of such buildings is sometimes difficult because of the frequent remodeling of the interiors of these establishments. The cause of the increase of black rats in residential areas from the mid-1990s is unknown, but I suppose that an increase of aged population living alone is one of the factors. Single families composed of people 65 years old or more tended to increase rapidly in the 1990s in Japan [20]. It is sometimes difficult for such people to rearrange rubbish and garbage around them to control rats [21, 22].

4. Norway rats in busy street

The removal system of garbage is different between big buildings and small buildings. Big buildings of 3,000 m^2 or more in total floors occupied by tenants such as stores, offices and hotels are regulated by the Building Standard Act. In these buildings, garbage is treated by building management companies. On the other hand, small buildings are out of regulation, and garbage in these buildings is usually kept in plastic bags and put along roadside (**Figure 3**). Accordingly, small buildings supply Norway rats with food sources, and the rats have become big problems in busy streets.

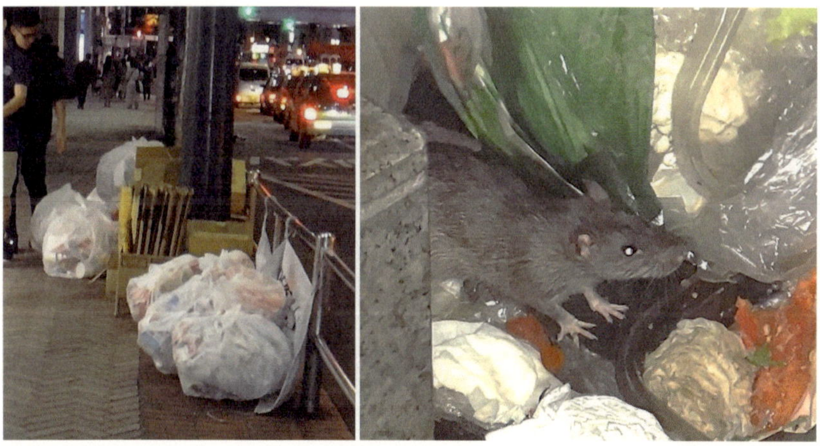

Figure 3.
Garbage in plastic bags along roadside, and a Norway rat attacking such bags.

5. Conservation of island ecosystem

Since the early-2000s eradication campaigns of rodents have been conducted in islands such as Yururi-Moyururi in Hokkaido, northern Japan, and the Ogasawara Islands, southern Japan [14, 15]. These projects were supported by the Ministry of Environment to conserve island ecosystems.

Author details

Tatsuo Yabe
Rat Control Consulting, Yamato, Japan

*Address all correspondence to: rccty@js8.so-net.ne.jp

References

[1] Bentley EW. A further loss of ground by *Rattus rattus* L. in the United Kingdom during 1956-1961. Journal of Animal Ecology. 1964; **33**: 71-373

[2] v Bülow B. Zur Verbreitung der Hausratte, *Rattus rattus* L., in Mittereuropa während der letzten Jahrzehnte. Zeitschrift für angewandte Zoologie. 1981; **68**: 67-94

[3] Lund M. Rodent problems in Europe. In: Prakash I, editor. Rodent Pest Management. Florida: CRC Press; 1988. pp. 29-33

[4] Lund M. Commensal rodents. In: Buckle AP, Smith RH, editors. Rodent Pests and their Control, 2nd Edition. Oxfordshire: CABI; 2015. pp. 19-32

[5] Anon. Black death-knell? Urban Wildlife News. 1991; **8** (2): 17

[6] Marsh RE. Rodent problems on the north American continent. In: Prakash I, editor. Rodent Pest Management. Florida: CRC Press; 1988. pp. 1-11

[7] Mabbett T. Commensal rodents–man's most versatile mammalian enemies. International Pest Control. 1992; **34**: 144

[8] Yabe T. *Rattus rattus* in urban buildings in Japan. International Pest Control. 1993; **35**: 96-97

[9] Yabe T. Probable relation of increase in big buildings and road coverage to the rise and decline of *Rattus rattus* in Sapporo, Japan. Medical Entomology and Zoology. 1997; **48**: 251-255

[10] Yabe T. Changes in species composition of the Norway rat (*Rattus norvegicus*) and the roof rat (*R. rattus*) in urban area (In Japanese with English abstract). Medical Entomology and Zoology. 1997; **48**: 285-294

[11] Yabe T. Urbanization and changes in species composition of commensal rats (In Japanese with English abstract). In: Matsuoka H, editor. Progress of Medical Entomology and Zoology. Tsu: Mie University Press; 2016. pp. 333-346

[12] Yabe T, Asai R, Takagi Y, Kohsaki K. Reason for the recent increase of roof rat infestation in residential areas in Tokyo. Medical Entomology and Zoology. 2000; **51**: 211-213

[13] Yabe T, Otomo T, Harashima T, Shigeoka H, Yamaguchi K. Breeding season in urban population of Norway rats in Yokohama estimated from age composition (In Japanese with English abstract). Medical Entomology and Zoology. 2016; **67**: 199-202

[14] Hashimoto T. Eradication and ecosystem impacts in the Ogasawara Island. In: Kawakami K, Okochi I, editors. Restoring the Oceanic Island Ecosystem: Impact and Management of Invasive Alien Species in the Bonin Islands. Tokyo: Springer; 2010. pp. 153-159. DOI: 10.1007/978-4-431-53859-_23

[15] Yabe T, Minato R, Hashimoto T. Breeding under snow cover in Norway rats (*Rattus norvegicus*) on uninhabited islands in Hokkaido, Japan. Russian Journal of Theriology. 2017; **16**: 43-46. DOI: 10.15298/rusjtheriol.16.1.04

[16] Ikuzawa M, Yonemoto S, Nishio Y. Changes in commensal rodents and their ectoparasites in Osaka City (In Japanese). Japanese Journal of Sanitary Zoology. 1973; **23**: 265

[17] Hata K. Changes in commensal rodents in Tokyo (In Japanese). Nezumi Joho. 1993; no. 44: 47-59

[18] Watanabe Y. Changes in commensal rodents in urban area (In Japanese). Kankyo-Eisei. 1979; **26** (1): 16-20

[19] Tanikawa T. Difference in consumption of warfarin baits and lethal days between resistant and susceptible colonies of the roof rats, *Rattus rattus*, in Japan (In Japanese with English abstract). Japanese Journal of Sanitary Zoology. 1991; **42**: 99-102

[20] Ministry of Internal Affairs and Communications. Whitepaper on Population Aging in 2017 Fiscal Year [Internet]. Available from: http://soumu.go.jp/h25/html/nc123110.html [Accessed: 22 December 2020]

[21] Motoki M, Izumi Y. Black rat control in a house of an aged man living alone (In Japanese). Nezumi Joho. 2015; no. 71: 1-5.

[22] Yaguchi N. Society of aged people and rat problems (In Japanese). Nezumi Joho. 2015; no. 72: 1-9.

Chapter 2

The Influence of Electromagnetic Fields on the Behavior of Mice

Roberto Carlos Vera and Israel Muñoz

Abstract

At the present, the development of life has led animals to have different behaviors in their evolutionary cycle, especially mice. For this reason, when mice are expose to physical agents such as electromagnetic fields, different behaviors can be found within their habitat and diet. Therefore, the analysis based on observation of the behavior of mice exposed to electromagnetic fields of different frequencies have been developed in the city of Potosí, Bolivia, which is located at an altitude of 3950 MASL. The methodology applied is the explanatory and longitudinal research. It is concluded that the influence of electromagnetic fields on the behavior of the mice generates a high stress index, influencing the change in the social behavior of the mice.

Keywords: electromagnetic fields, mice, mouse stress, mobile phones, mouse hyperactivity

1. Introduction

At the end of the twentieth century, humanity had great scientific advances, one of the main ones being the development of telecommunications. Today they have become particularly important since these technologies have allowed globalization and thereby improve both life in general and business life. Year after year, telecommunications technology evolves and with it, different applications have been created, which spread rapidly in all regions of the planet. Together with this communication phenomenon, electromagnetic signals, which have a certain influence on the environment, have increased due to the power of electromagnetic irradiation. This is how the electromagnetic spectrum for its irradiation depends on its wavelength and its frequency, concentrating a certain amount of propagated energy in the form of packages called quanta, postulated by Max Planck (1858–1947).

$$E = h\nu \tag{1}$$

where h is the Planck's constant (6.626×10^{-34} J × s^{-1}) and ν is the wave frequency (Hz).

In consideration of this Planck's postulate, the propagation of the energy of the irradiation of the electromagnetic field has an influence of interaction with matter; in addition, the living organisms of the planet are considered to be biochemical and bioelectric, which adapt to different conditions of the environment where it is evident that today pollution has increased compared to previous times. For this reason, the propagation analysis of electromagnetic waves must be based on the optical properties of interaction with matter that, depending on the distance to the source and the time of exposure, more electromagnetic energy can be concentrated.

Consequently, it is important to consider that the electromagnetic force is composed of electric and magnetic fields, which are intrinsic properties of matter and can be presented statically and/or dynamically, where the emission of these variables is known as "electromagnetic radiation." The moving electric charges produce electric currents of different intensity, giving rise to the propagation of electromagnetic waves in the medium. **Figure 1** shows the schematic distribution of the electromagnetic spectrum.

It is important to note that electromagnetic waves are transverse, between the intensities of the magnetic field E and the magnetic field H, generating an irradiation called the Poynting S vector, in honor of John Henry Poynting [1].

$$S = E \times H \tag{2}$$

Electromagnetic radiation does not need a medium to propagate; however, air is known as a propagation medium which has certain conduction impedance, defined by the following equation:

$$Z = \sqrt{\frac{K\mu_0}{K\varepsilon_0}} = 377[\Omega] \tag{3}$$

where Z is the air impedance, μ_0 is the magnetic permeability, ε_0 is the electric permittivity, and K is the dielectric constant. According to Milford et al., K is 1.00059 [2].

The air impedance must be quasi-constant. Additionally, the interaction of electromagnetic fields with matter must be considered, where different behaviors are expected, which is due to the different electromagnetic optical properties. According to Kraus and Fleisch, other properties of electromagnetic radiation emission are attenuation, which consists of the interaction of the electromagnetic

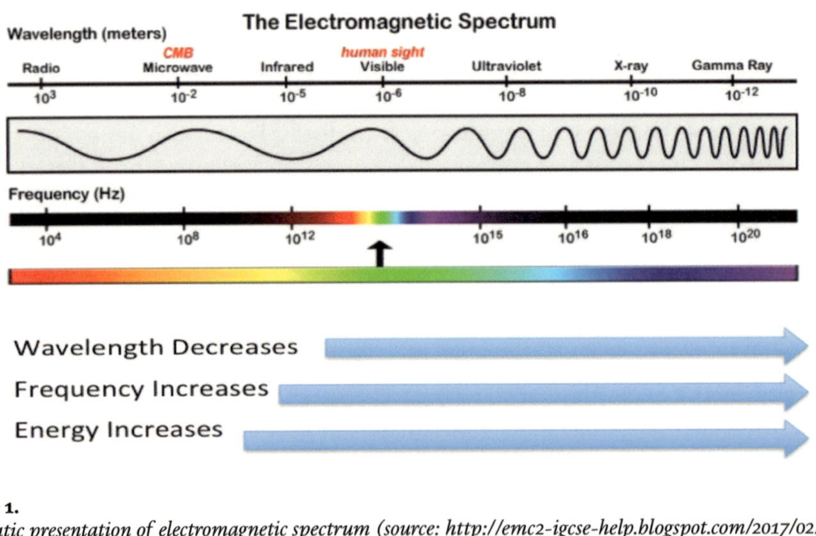

Figure 1.
Schematic presentation of electromagnetic spectrum (source: http://emc2-igcse-help.blogspot.com/2017/02/311-identify-order-of-electromagnetic.html).

field with some material, which has the property of exponentially decreasing its initial value and is given by the following equation:

$$\propto = \sqrt{\frac{\omega\mu\sigma}{2}} \tag{4}$$

where ∝ is the attenuation constant (Np × m^{-1}), μ is the permeability of the medium (H × m^{-1}), σ is the conductivity of the medium (Ω^{-1} × m^{-1}), and ω is the frequency (rad × s^{-1}).

The wave penetration consists of a damping of the incident waves, since it is inversely proportional to the thickness of the material, which dissipates its energy as it travels. This energy is transformed into heat, which occurs by incidence of the wave that when crossing causes the molecules to vibrate causing molecular movements [3]. It is given by the following equation:

$$\delta = \frac{1}{\sqrt{\pi\mu\sigma}} \tag{5}$$

where δ is the penetration constant (m), μ is the permeability of the medium (H × m^{-1}), σ is the conductivity of the medium (Ω^{-1} × m^{-1}), and π is number pi.

It is noted that these forms of communication have reached the general public, thanks to the constant innovations in systems and infrastructures introduced by mobile phone companies. Despite the fact that a large part of the current terminals belong to the second generation of mobile telephony, the current and future new developments are focused on the evolution of the third and even the fourth generation (3.5G, 3.75G, and 4G) [4]. However, we must consider in the last 2 years, there has been a great technological advance in 5G technology. Although it is still under development and testing, it will be implemented very soon (**Figure 2**).

Given this description, it clarifies that information and communication technology (ICT) is applied today, which facilitate new roles to work efficiently. For this reason, the work carried out by Ruiz-Palmeros and his collaborators in this technology concludes with the following context: "The first factor, excessive or inappropriate use of the mobile phone, included the difficulty in controlling behavior and impulses. The second factor was abstinence and the grouped elements in which concern was expressed about the possibility of not having a telephone. The third factor, by elements, referred to the difficulty of stopping the use of the telephone and family problems. The fourth factor explains the increase in data consumption" [5]. This situation leads to raise a critical and reflective

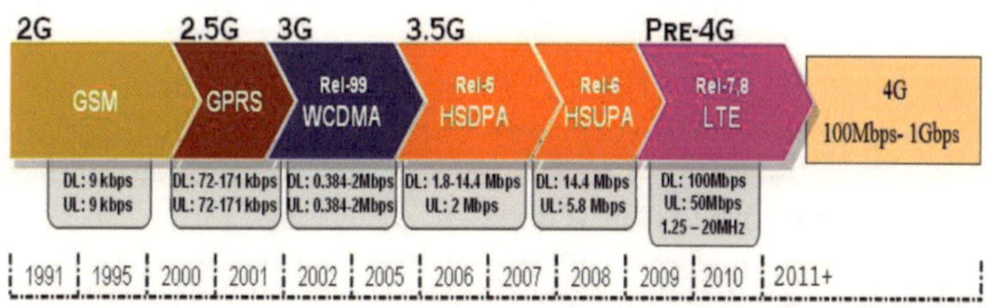

Figure 2.
Cell phone evolution timeline (source: https://cienciaysociedad/documentos/doc/El_Telefono_Movil_ETSIT-UPM.pdf).

aspect of the gravity or influence that communication technology has on biological bodies; so, the interest of this work focuses on observing the behavior of the direct and indirect influence of the emission of electromagnetic fields with a thorough analysis of experimentation in laboratory organisms, such as mice.

Mice are one of the most fertile and numerous groups of mammals on Earth due to the extraordinary ability of their populations to reproduce. They have large numbers of offspring, which is one of the primary reasons they make up the largest group of mammals; the second is that they have a short gestation (pregnancy) period. They are grouped in the Rodentia order and are characterized by very sharp and curved teeth like a chisel that is used to gnaw hard objects.

Rodents are the order with the most species within the group of mammals; there are more than 400 genera and some 2000 species. Among the best known species are mice, rats, chinchillas, and squirrels. Rodents that are closely related to humans (commensal rodents) such as the brown, Norwegian, or water rat (*Rattus norvegicus*), the black or roof rat (*Rattus rattus*), and the house mouse (*Mus domesticus* and *M. musculus*), which have spread throughout the world, taking advantage of their simple body designs, a high reproductive rate, a general diet, and a sophisticated behavior pattern that allowed them to avoid the most cunning attempts at eradication. Knowledge of the response of rodents to their environment can help explain their behavior patterns and allow us to propose or establish control methods [6]. Based on these considerations, it is important to note Mary Johnson's manuscript: "*Genetically defined and genetically modified mice and rats are widely used in research to analyze the function of specific genes, and to serve as experimental models for different human diseases. Thousands of these strains are available, with infinity of genetic alterations, selection of the strain and several applications*" [7].

Given these described aspects, the interest of this document focuses on verifying in a qualitative and quantitative way the real behavior of the exposure of mice to non-ionizing electromagnetic fields. Additionally, it is important to take the main biological aspects of mice according to the laboratory animal management guide. These groups mark their vital areas with urine, forming a network of odors that then allows them to overcome narrow bridges in total darkness. Dominant males and reproductive females create 3-cm-high olfactory stalagmites that announce the presence of their territories to animals nearby. In addition, it is important to take note of the sensory characteristics that they have, which are as follows:

- Smell: used to guide their movements around vital areas where feces, urine, and genital secretions contribute to leaving traces of odor. Traces are detected and can be followed or avoided by other individuals.

- Touch: their main action is the whiskers; this is because the mice and rats have the ability to control their position; they are in constant movement during the exploration, contacting the ground, walls, and any nearby object; in addition, they are the ones that announce the changes of the climate in the environment.

- Hearing: they have a keen sense of hearing, in addition to listening in the range audible to humans, they can capture ultrasonic sounds including those emitted by themselves in that range (between 22 kHz and 90 kHz), and these are used for social communications between them.

- Sight: they are specialized for night vision; they have high sensitivity to light, but poor visual acuity. Rodents have good depth perception and are able to correctly assess the effort required to perform any type of jump; apparently they do not detect colors, capturing them as gray variants; yellow and green are probably the most attractive colors being perceived as light gray.

- Taste: the sense of taste is highly developed; they have a great ability to detect minimum amounts of bitter, acidic, toxic, or unpleasant substances, which complicates control with toxic baits.

It is also important to consider the predominant physical abilities: digging, climbing, jumping, and mainly gnawing, where their upper and lower incisors constantly grow, being worn away by this action [6]. This is how these conditions set the characteristics for experimentation and development of executed work.

2. Materials and methods

Based on the statements described above, it is important to consider the main behavioral characteristics that developed when mice exposed to electromagnetic fields. For this situation, it is important to reflect on the geographical place where it was made, the city of Potosí, with an average altitude of 3950 m above sea level, with temperatures ranging from 18°C in spring and summer to less than 5°C in the autumn and winter seasons.

The procedure carried out in this quasi-experimental evaluation starts from an exhaustive analysis in the control and observation of mice before and after irradiation to electromagnetic fields generated by mobile phones and signals that exist in the environment. For this reason, the evaluation is carried out in stages or phases, the selection of samples in an organized and equitable manner in each glass box. The first phase consisting of the time of adaptation and acclimatization to the environmental conditions of the region; in this, the metabolic data that the mouse undergoes and its behavior within its habit are registered before being affected by electromagnetic fields, learn about the different abilities particular of mice. Also, at this stage, it is important to record the levels of non-ionizing radiations, for its analysis and to subsequently observe the increase that is made with mobile sources such as cellular sources. In the second phase, it analyzes the characteristics of the cell phones, where the irradiation power must not exceed the permissible limits established in the ICNIRP. Subsequently, the phones are introduced gradually, observing and recording the behavior of the mice within their habit, applying the Likert scale method to demonstrate the level of behavior that each mouse has and thus, demonstrate the influence caused by the fields electromagnetic on mice.

Thus, the mice are housed in glass boxes with appropriate dimensions which are designed to facilitate their behavior since they:

- have adequate, safe space and protect them from external threats;
- provide adequate ventilation;
- provide ease of cleaning and are resistant to frequent disinfection and sterilization;
- allow the observation of the animal;
- facilitate the access or extraction of animals to verify their increase or decrease in body mass;

- there are no sharp edges or projections that can cause injury to the animal; and
- a running wheel is incorporated for their development.

In consideration of these aspects, it is also important to consider the instruments for analyzing the irradiation of electromagnetic fields. For this, three radiofrequency spectrum analyzers were used to observe the level of concentration of concentrated electromagnetic energy in the environment where the rats are located. These instruments are Spectran NF5030, HF2025E, and HF6040, from the German industry, with calibration certification; they analyze electromagnetic pollution with a margin of error of 0.7%, at 2.5% based on the standards established by the ICNIRP [8]. These regulations allow us to establish the real situation in a quasi-experimental environment, where the interest is focused on observing the behavior of animals exposed to these electromagnetic fields that are not noticeable.

With all these aspects, a rodent laboratory (bioterium) was implemented, considering the main aspects of hygiene, safety, and size control in the intrinsic metabolism of each mouse. It is also important to mention that the biosecurity elements are particularly important, where they can ensure control against any pest and/or disease that may be emitted while cleaning or controlling the dwellings where the experimental mice are located (**Figure 3**).

Once the environment is adequate, we proceed with the applied research methodological principles [9], seeking to generate knowledge on the topic developed. Therefore, this is fundamentally based on the findings that the use of wireless communication technology used in our environment implies addressing the issue of the current situation of settlement and/or proliferation of telecommunication antennas. For this reason, the study of this work concentrates on an explanatory methodology; it is used in order to try to determine the causes and consequences of the aforementioned phenomenon, giving the survey of why in some regions of the planet, there is so much problem of settlement or exposure to electromagnetic fields that are generated by the antennas of telecommunication and/or by the latest generation cell phones based on a coherent state of the question and without manipulation of information. We use of the logical tools of research centered on the inductive method this is focused on the observation and analysis of the situation, allowing conclusions to be drawn from the events that occurred in mice exposed to electromagnetic fields, considering that this is quasi-experimental, in the fact that it is intended to manipulate some specific variables, such as the density of mobile phones, which emit a certain amount of electromagnetic radiation, taking into account that there is no full control over all variables. Finally, longitudinal monitoring is assumed, which characterized the monitoring of the behavior of mice exposed to electromagnetic fields considering a specific observation period (**Figure 4**).

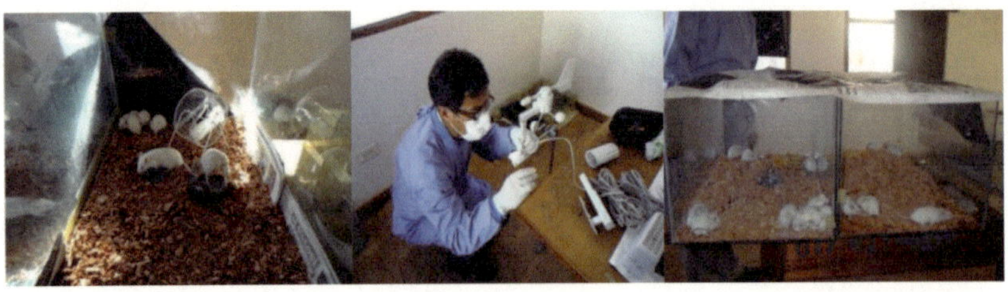

Figure 3.
Implementation of the Mice Lab at the Physics Department, "Tomás Frías Autonomous University."

Figure 4.
Laboratory instrumentation (bioterium), Physic Department, "Tomás Frías Autonomous University."

3. Results and discussions

Based on the established methodologies, we start with the climatic variables of the laboratory such as temperature and relative humidity (RH) in the environment. For this cold weather situation, the mice gradually adapt according to their habit of developing in glass cages, where they present concerns for their immunological and psychological development. Initially, the mouse has a social characteristic and is kept in groups without any problem; these groups form quickly once they are introduced to the glass cage. However, the males of both strains (boxes) begin to show their aggressiveness on the 15th day, even though these groups have not fully established themselves in their habitat. Low temperatures in the laboratory cause a death of 15% of the samples, which causes a controlled heat system to be introduced at certain hours of the night, in order to avoid the loss of the samples. Once these environmental conditions of temperature and humidity were established, the mice showed greater social activity among themselves. At the time of providing the corresponding food, there should have been a procedure for an adequate food balance, such as composition, meeting growth needs, and coat maintenance, the latter being a main aspect of observation, which thanks to this I show some allergies and/or poor digestion in gnawing some cereals.

Therefore, the feeding that is supplied to the mice must have the necessary amounts in fiber and nutrients; in addition, it must be considered that these animals always seek to gnaw some food; that is why much of the diet is concentrated in cereals such as wheat and corn. However, their diet is also concentrated on green foods and nuts in addition to proportions of potatoes, carrots, and other foods that help in growth. It is evident that balanced food that exists for domestic animals such as cats and dogs are attractive to the mouse, the same that causes the fur of the mouse to increase and be much finer. This leads us to have two affirmations: the first is that the food has enough vitamins and they consume in greater quantity in addition to having the corresponding hardness to gnaw. The second is that the taste sensation in some mice causes the bowel movements to be inconsistent but rather causes foul smelling diarrhea that is not favorable to the mouse. Therefore, the nutrition of the mice had processes through which the biological body transforms and uses the nutrients to obtain enough energy, as well as to maintain and repair the tissues since the organism needs to acquire an external contribution of matter, essential for getting the substances that regulate the metabolic processes of the mouse.

It is important to consider the feeding for cold places should be a maximum of 5.7 to 7.5 g of food per mouse; the above is subject to consideration depending on the climate of the region. It should also be considered that the water supply should never be missing in either of the two sample boxes, in a quantity of 250 ml per day. The adaptation time of the mice in the climatic conditions of the city of Potosí

Box 1			Box 2		
Mouse ID	Initial mass (g)	Final mass at the end of adaptation time (g)	Mouse ID	Initial mass (g)	Final mass at the end of adaptation time (g)
1	11.54	19.45	1	10.45	20.71
2	13.45	20.45	2	9.25	—
3	11.23	23.81	3	10.33	24.56
4	10.45	—	4	12.31	24.67
5	12.25	29.83	5	11.37	20.12
6	12.64	29.68	6	12.15	31.84
7	11.34	—	7	11.57	22.64
8	10.57	—	8	10.28	—
9	12.52	24.67	9	9.92	27.37
10	9.89	23.58	10	9.72	—
11	10.29	—	11	10.84	25.36
12	11.64	23.37	12	11.71	23.37

Source: U.A. "T.F."—FAUTAPO—Physic Department, Investigation Roberto Vera.

Table 1.
Body mass control at adaptation time.

was 33 days; currently, there was the loss of samples due to the climate conditions, where the extreme coldest temperature was -4°C. Despite the fact that the environment is controlled, it is important to consider that carbohydrate feeding increases between 10 and 15% on these cold days, since the body of the mice needs to create a greater amount of fat for their protection from weather conditions. In the laboratory, there is a heat regulator so that the temperature does not drop abruptly in the implemented environment (**Table 1**).

Once the adaptation of the mice has been achieved, parallel to this and for a time of 70 days, the accumulated average values of the power density of electromagnetic radiation are shown; according to the density of users, these levels of power increase during daylight hours, especially in the periods from 12:00 to 13:00 and approximately 18:30 to 20:00. This action is because many people today use the mobile phone, for immediate communication, which causes the concentration of power in some parts of the city to increase; so it is right where there is a telecommunication antenna, especially mobile telephony, will have a quasi-similar behavior to that of the following graph.

Although the graph in **Figure 5** shows high peaks, it must be considered that the irradiation activity was continuous; this means that the levels of electromagnetic radiation increased progressively, making use of commercial mobile phones. Where the power density generated by each of the cell phones does not exceed the value of 5 (mW/cm^2), this is based on the ICNIRP international standards, giving certainty of compliance with this.

In knowledge of these aspects, it is important to explain the behavior of mice exposed to these electromagnetic fields. It should be considered that the irradiation was progressive, considering that adaptation to the environmental conditions of the city of Potosí, also influence mice, which influences the behavior of hyperactivity, feeding, and growth. For the analysis of the irradiation of electromagnetic fields on the mice, a detailed control was carried out, starting with the division and marking of each mouse, with great care, where it identifies individually the behavior of feeding (appetite), thirst (water consumption), sleep, aggressiveness, hyperactivity, irritability of the eyes, and bedding (mouse nest). Each of these aspects is considered as a variable, the same that is registered in a Likert table; this evaluation allows describing the physiological conditions, that is, the stress level of each mouse in the adaptation period, as well as in the gradual exposure to concentrated electromagnetic field levels in the experimental environment. The Likert levels adopted for this situation were as follows: normal level: 1 to 21; medium or moderate level: 22 to 42; and high or acute level: 43 to 63. Based on these considerations, **Figure 6** shows the behavior of the mice in general.

These last two graphs show admirable considerations at the level of the behavior of mice, especially on stress, generated by electromagnetic fields. It is interesting to discuss why this exposure to electromagnetic fields influences the behavior of mice, taking the following explanatory points of observation that are evident in experimentation.

Adaptation aspect and without exposure to electromagnetic fields: mice in the adaptation period show conflicting behavior in the first week, and in some cases, the isolation of some of the samples is evident, causing the normal level of stress to be high; however, over time, this decreased in this period. Despite having an extensive controlled diet of fibers, minerals, and others, some of the mice were unable to adapt to the environment; this is due to the low temperatures recorded. The death of the mice was not only due to the climate but also due to the fight between the dominant males of the herd; there was also the fight between some females and some males; this is due to the desire for reproduction which produced the death of some samples. Between the second and fourth weeks, it is evident that the mice begin to have an organizational system with respect to their habitat. From the third week on,

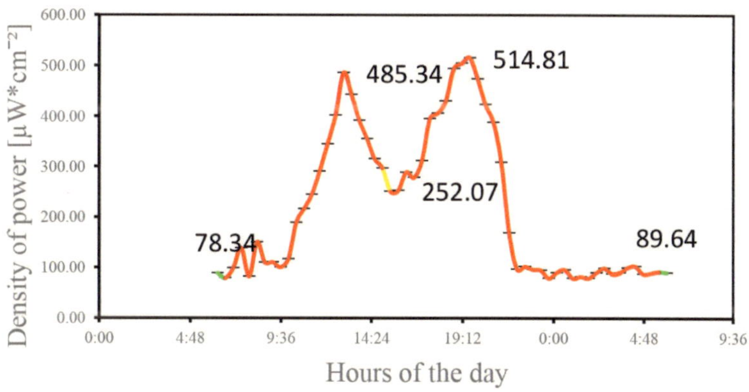

Figure 5.
Temporal behavior of the power density of non-ionizing radiation.

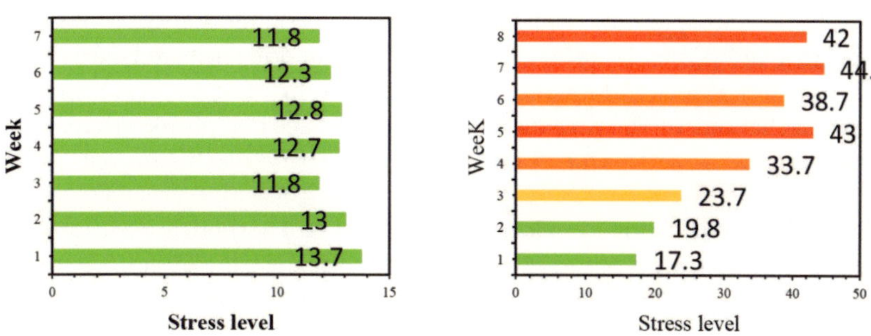

Figure 6.
Chart of stress levels without exposure (left) and with exposure (right) to EMF.

the mice show permanent exploration activity within their vital areas or domains they constantly explore in their environment, both the known and new elements or objects, whether sniffing, gnawing (biting), and tasting food and liquids found in the glass cage. We consider the organization and the exploratory activities carried out by the mice as relevant, however it is important to highlight that, at the time of periodically cleaning the cages, it is evident that urine stools defecation, frequently they are positioned in the same place away from the power supply and water.

Aspect of behavior of mice with exposure to electromagnetic fields: it is evident that in the second graph (right) of **Figure 6**, it shows that the behavior of mice in their stress levels gradually increases, from the lowest (tranquility), going through the moderate level to the level of watery aggressiveness. It is evident that the influence of electromagnetic field or radiation has a great direct influence on the behavior of mice exposed to these physical agents; despite the fact that the average energy accumulated per minute does not exceed 350 mJ, it is the same as it is a very low energy, compared to the molecular resonance that occurs for the ionization of the molecules. However, considering the physical principles of electromagnetic wave interaction with matter and its penetration on these biological bodies, we affirm that energy can accumulate according to the principle of wave superposition for certain time. This aspect causes an invisible concentrated field of energy to exist in the cages and the experimental environment, the same that increases the temperature level that is perceptible to mice; so, the behavioral changes in their habitat and physiology are very noticeable and progressive. The influence of electromagnetic radiation, emanating from different sources, causes a natural and necessary response for survival in the mouse; so, these actions give rise to the following aspects:

- In their habitat: a disordering in the path of their health needs, compared to the situation adaptive clutter in your shelter is more evident, foods are scattered everywhere. The presence of mobile phones in operation (without sound, vibration, or light emission) triggers a restless response in mice, where they try to gnaw and tear the object with greater force; in addition, they work on more than two subjects, trying to hide the equipment; for this action, they use the parts and other extremities, carrying small debris to cover said object, even though the mice avoid approaching the point where the mobile terminal is located. This aspect not only occurs with the cell phone but also with other electrical and electronic instruments, such as coils, radios, and current extenders.

- In their diet

- In the liquid consumption: they present a higher consumption which causes some fights and aggressiveness between them, noticing that we provide more water to avoid fights. Despite dividing the proportions of water into three different containers, the aggressiveness persists. In addition, it is important to mention that one of the water sources is close to the cell phone where the mice avoid going to it, with some of the samples approaching this source.

- Within their hyperactivity: a low performance is observed, since there is greater aggressiveness among the mice, which at the time of getting on the spinning wheel, some of them approach and push it, causing the fall of this.

- In their physiology: within the last 3 weeks of irradiation with electromagnetic fields, the mice have great irritation in the eyes, which when approaching to the activated source of electromagnetic radiation (mobile phone) avoid being close to it. The mice feel discomfort in the whiskers more frequently when the active electromagnetic signal is increased (when the cell phone is switched on); the frictions that are made in the presence of this physical agent develop irritation in the mouse's snout.

All these exposed aspects show an explanation of how the influence of electromagnetic fields in prolonged periods influence the behavior of mice both in their stress levels and in their physiology.

4. Conclusions

The presence of electromagnetic fields within the environmental environment has grown progressively according to the human population density. The direct or indirect influence of these fields generates an uncomfortable presence in the habitat of an animal, especially for those animals that are sensitive to the variation of the intensities of static or dynamic electromagnetic fields, proof of this is the behavior of the mice within this study which are sensitive to the increase in the levels of electromagnetic fields.

Mice, without the influence of electromagnetic fields, present an organization within their habitat, which demonstrate dynamic hyperactivity in their life cycle. On the other hand, if the conditions of their habitat are abiotic due to the influence of electromagnetic fields, the mice present different levels of behavior, raising their stress conditions, for example, in aggressiveness and irritation of the eyes.

In an epilog, it can be concluded that the action of the electromagnetic fields generated by mobile phones directly influences mice on their level of behavior and their habitat. Furthermore, when the presence of this signal exists, the mice avoid proximity by activating intrinsic prevention in each one of them.

Acknowledgements

The authors are grateful to the FAUTAPO foundation, for making possible the purchase of equipment for the investigation of electromagnetic fields in the non-ionizing spectrum.

Author details

Roberto Carlos Vera[1*] and Israel Muñoz[2]

1 UTEPSA University, Santa Cruz, Bolivia

2 National Polytechnic Institute, Mexico City, Mexico

*Address all correspondence to: robertrormc@gmail.com

References

[1] Hecht E, Zając A. Optics. New York: Addison-Wesley; 1974

[2] Reitz JR, Milford FK, Christy RW. Fundamentos de la Teoría Electromagnética. México D.F.-México: Addison-Wesley Iberoamericana; 1986

[3] Fleisch y DA, Kraus JD. Electromagnetics: With Applications, New York: McGraw-Hill; 1999

[4] Sánchez FR, Vacas FS, Hervás JMG. El Teléfono Móvil, Producto Estelar de la Red Universal Digital. Madrid: ColorIURIS; 2010

[5] Ruiz-Palmero J, Sánchez-Rodrígue J, Trujillo-Torres JM. Utilización de Internet y dependencia a teléfonos móviles en adolescentes. Revista Latinoamericana de Ciencias Sociales, Niñez y Juventud. 2016;**2**(n° 14): 1357-1369

[6] Ministerio de Salud. Guia de Manego y Cuidado de Animales de Laboratorio: Ratón. Lima: FIRMAT S.A.C; 2008

[7] Johnson M. Mater Methods. Princeton, New Jersey: Synatom Research; 2012

[8] International Commission On Non-Ionizing Radiation Protection. Guidelines for Limiting Exposure to Time-Varying Electric, Magnetic, and Electromagnetic Fields (up to 300 GHz). Munich: ICNIRP publication; 1998

[9] Rivero DSB. Metodología de la Investigación. Madrid: Shalom; 2008

Chapter 3

Effect of Diet and Water Availability on *Rattus norvegicus* (Rodentia: Muridae) Distribution

Tatsuo Yabe

Abstract

The distribution of the Norway rat *Rattus norvegicus* extends from the subarctic to the subtropics in Japan; yet it is limited by several factors. I discuss appropriate diet, water balance, and temperature as limiting factors based on surveys in the subarctic zone (Yururi-Moyururi, uninhabited islands in Hokkaido), the temperate zone (a business district in Yokohama and an uninhabited islet, Kaiho-2 in Tokyo Bay), and the subtropics (the Hahajima Islands in the Ogasawara Archipelago) in Japan. In Yururi-Moyururi, the rats recruited new generations in their population not only in the summer but also under snow cover, probably by preying on carcasses of their own species. In Yokohama, peaks of recruitment of their new generations were found in the winter and the summer, though the season with peaks changed every year. In Kaiho-2, rats stopped recruiting in the winter because of dehydration, and over the winter the group lost body mass as a result of body fat consumption. In Hahajima, rats lost body mass and preyed mainly on plant matter because of chronic dehydration. I conclude that protein-rich diets and water balance, but not temperature, are basic factors in the distribution of the Norway rat.

Keywords: *Rattus norvegicus*, geographical distribution, limiting factor, protein-rich diet, water balance

1. Introduction

The Norway rat *Rattus norvegicus* Berkenhout is one of the commensal rodents, along with the roof rat *R. rattus* Linnaeus, the Polynesian rat *R. exulans* Peale, and the house mouse *Mus musculus* Linnaeus. These rodents expanded their distribution worldwide by taking advantage of human activities [1–3]. However, they have limitations in their geographical distributions. Brooks and Rowe [2] point out that Norway rats are fundamentally fitted to the temperate zone, and they are less prosperous in tropical and subtropical climate zones, whereas roof rats thrive in tropical and subtropical climate zones. The question arises as to whether Norway rats are fitted to the temperate zone due to the mild temperature. Tomich [4] points out that mild temperature is secondary to appropriate diet as a factor in determining the Norway rat distribution. Although they are omnivorous, Norway rats require a diet containing a certain amount of animal matter or that is protein-rich [5–7].

Many species of seabirds nesting on or near the ground or in burrows are vulnerable to predation by Norway rats because of the terrestrial behavior of the rats [1, 8]. Rats on an island in the Aleutians were supposed to prey on seabirds and to restrict the productivity of shorebirds and land birds by preying on the birds' food [9]. For Norway rats, such subarctic and subantarctic zones are severe environments in the cold season; when the rats' reproductive activities are depressed, their ears, legs, and tails are frostbitten, and their mortality rate is higher [10, 11]. However, Yabe et al. [12] discovered Norway rats breeding under snow cover on uninhabited subarctic islands in Japan. This fact suggests that they breed even during the cold season or under snow cover when an appropriate diet is available.

Also, the tropical and subtropical climate zones seem to be severe environments for Norway rats. Norway rats in the tropical climate zone are distributed in patches in limited areas such as seaports, irrigated villages, and large cities [2, 13–15]. Yabe et al. [7] found that the body mass of Norway rats on islands in the subtropical climate zone was smaller than those in the other habitats in the subarctic climate zone and the temperate climate zone in Japan because of a protein deficiency. Norway rats on the islands in the subtropical climate zone preferred plant matter to animal matter. On the other hand, Norway rats on an artificial islet in the temperate climate zone stopped breeding and lost body mass in the dry winter even though they preyed on some animal matter [16, 17]. Therefore, it seems that the appropriate diet changes depending on the habitat, and protein-rich diets do not always help Norway rats to thrive. Then commenting on the review by Yabe [18], I discuss the factors that cause the appropriate diet for Norway rats to shift based on their habitat and thus limit their geographical distributions.

2. Breeding in the subarctic zone

2.1 Breeding under snow cover

Yururi (168 ha, 43° 12′ N, 145° 35′ E) and Moyururi (31 ha, 43° 13′ N, 145° 36′ E) (referred to as Yururi-Moyururi hereafter) are uninhabited islands situated 2.5 and 3.7 km off the Nemuro Peninsula of Hokkaido, respectively (**Figure 1**). They are in the subarctic climate zone and have a mean annual temperature of 6.3°C. Both islands are flat and covered with low vegetation such as alpine plants and the bamboo grass *Sasa nipponica* Makino and Sibata. According to the local people, Norway rats intruded into these islands in the 1960s or 1970s from a boat used for fishing or light house construction.

Generally, the reproductive activities of Norway rats in the subarctic and subantarctic zones seem to be restricted in the summer. Schiller [11] found that the breeding season of Norway rats in a business district and in dumping sites in Nome in Alaska occurred exclusively in the summer. Pye and Bonner [10] also found that the breeding season of Norway rats in a coastal area on South Georgia Island in the subantarctic climate zone was in the summer from December to February. The most active breeding season for Norway rats on Yururi-Moyururi also seemed to be in the summer. Here, 63 (86.3%) of the 73 rats caught in late July and early August 2013 were born from June to July. However, 10 (13.7%) of them were born from December to March, the heavy snow season (**Table 1**) [12].

Data collected by a metrological station at Nemuro, a city close to these islands, show that the amounts of snowfall were 52, 41, 52, and 29 cm in December 2012 and January, February, and March 2013, respectively. No rats entered these islands in these years because boats are restricted from approaching these islands, and there were no wrecked vessels after these islands were appointed to be a sanctuary for birds in 2011. The distance from the Nemuro Peninsula to Yururi-Moyururi is over 1 km that is pointed out by Russell et al. [20] as a possible distance for Norway rats to swim. Therefore, the 10 rats on Yururi-Moyururi must have been born during

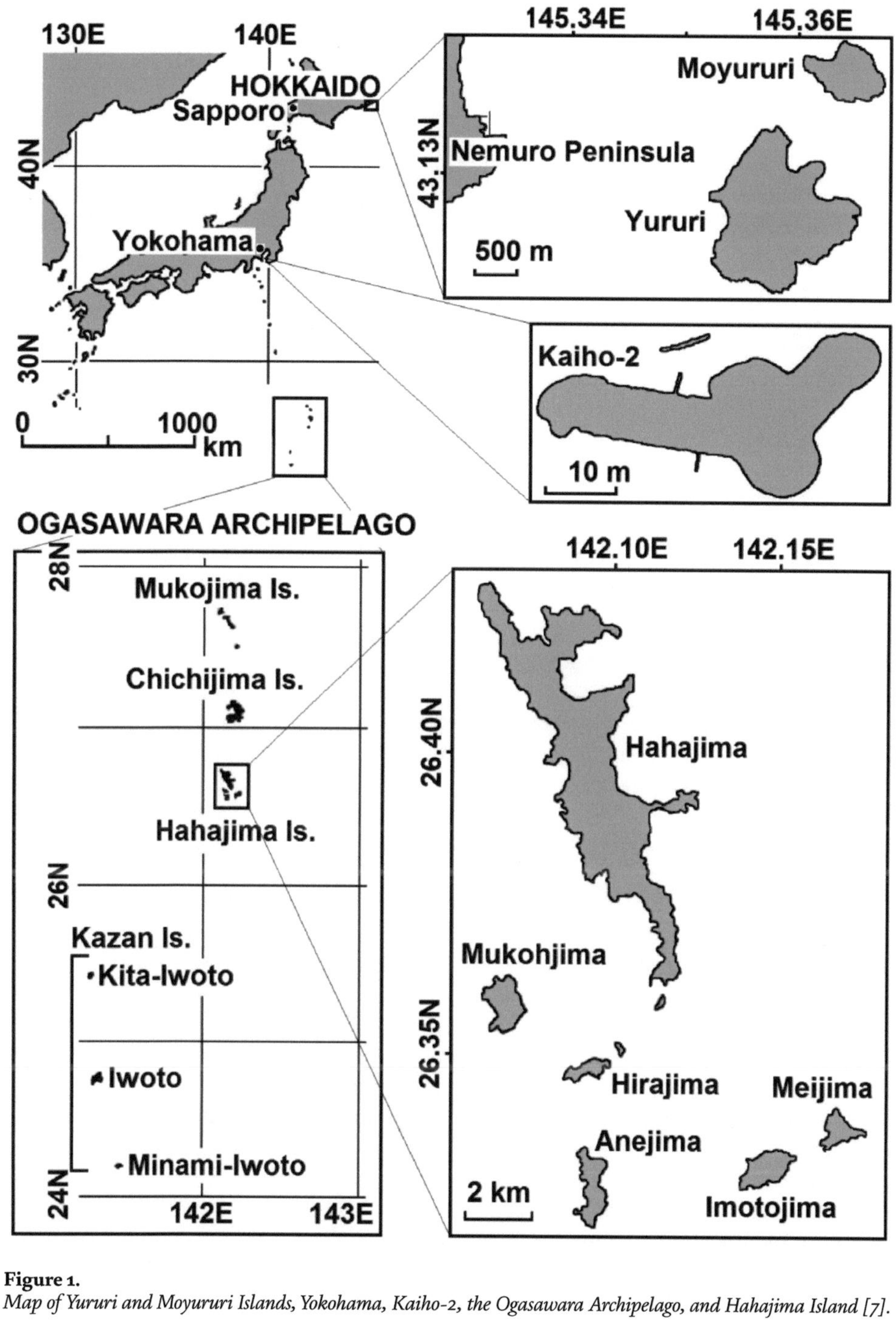

Figure 1.
Map of Yururi and Moyururi Islands, Yokohama, Kaiho-2, the Ogasawara Archipelago, and Hahajima Island [7].

the heavy snow season. Snow cover protects Norway rats from cold temperatures. The temperature at the ground level under 50 cm of snow cover, for example, is kept above −5°C, even when the air temperature is below −30°C [21]. Inukai [22] also showed that the temperature at the ground level under 1 m of snow cover was from 0 to −2.8°C when the air temperature ranged from −6 to −13°C in Sapporo, Hokkaido. Furthermore, deep snow cover stabilizes the temperature under the snow [23], and thus, snow cover likely provides comfortable breeding conditions for Norway rats. Maeda [24] also found evidence of the breeding of Norway rats under snow cover just after the melting of the snow in a forested area in Sapporo.

Age in months	Birth month	Number of rats			
		Male	Female	Total	Pregnant
1 (1.0–1.9)	July	28	24	52	7*
2 (2.0–2.9)s	June	10	1	11	1
3 (3.0–3.9)	May	0	0	0	0
4 (4.0–4.9)	April	0	0	0	0
5 (5.0–5.9)	March	3	0	3	0
6 (6.0–6.9)	February	1	3	4	2
7 (7.0–7.9)	January	0	2	2	1
8 (8.0–8.9)	December	1	0	1	0
Total		43	30	73	11

**1.2–1.7 months old.*
Seven females less than 2 months old were pregnant. Ages were estimated using a formula [12, 19] based on eye-lens weight. Birth month was calculated by subtracting the age from the date when the rat was caught. New data on pregnant females were added [12].

Table 1.
Age composition of Norway rats caught in late July to early August 2013 in Yururi-Moyururi.

2.2 Appropriate diet under snow cover

Norway rats on Yururi-Moyururi thrived and reproduced under snow cover without depending on human beings for their diet. In the case of Maeda [24], Norway rats ate mainly bamboo seeds and rodents such as gray red-backed voles *Myodes rufocanus* Sundevall, the population of which exploded after the bamboo-grass flowering. The voles usually make their nests in tunnels under ground, but in the snow season they make their nests and breed in the space under bamboo grass covered by snow [25, 26]. Therefore, it is likely that Norway rats could easily find and prey on such voles. However, there were no rodents or other small mammals except Norway rats on Yururi-Moyururi (T. Hashimoto, pers. comm.).

Birds of prey such as common buzzards *Buteo Buteo japonicus* Temminck and Schlegel are known to live on Yururi-Moyururi [27, 28]. It is likely that shallow snow and dead grass cover these islands at the end of autumn or beginning of winter. Rats running across such white snow are vulnerable to birds of prey [29], and rats running on dead grass may also be. Among birds of prey, common buzzards are known to feed on Norway rats [30], and they probably leave behind body parts of the prey as in the case of roof rats (**Figure 2**). The dense population of rats that were born during the summer will provide the necrophagous rats with many rat remnants as a food supply to winter and breed under snow cover. All the rats in Yururi-Moyururi were less than 9 months old (**Table 2**). This suggests that their life spans were shorter than in any other habitats such as the Hahajima Islands, a business district in Yokohama and an islet (Kaiho-2) in Tokyo Bay. Norway rats that were 13 months old or older were common in the latter three habitats (**Table 2**). Predation by birds of prey was probably one of the causes of their short life span. Snow cover in the heavy snow season protected Norway rats from such predators and helped them to breed during the winter.

2.3 Appropriate diets for breeding in summer

Meehan [31] reported that Norway rats become sexually mature at 2–3 months old, but it has also been found that they can become mature at less than 2 months old

Figure 2.
Remains of roof rats Rattus rattus *left by common buzzards* Buteo buteo japonicus *in the Ogasawara Archipelago (A: by T. Yabe; B: by F. Nomura, provided by PREC Inst. Inc.).*

	Age in months				
Locality	**<13**	**≥13**	**Total**	**% of ≥13**	**Reference**
Hahajima Islands	32	42	74	56.8	[7]
Yururi-Moyururi	73	0	73	0.0	[7, 12]
Yokohama	97	20	117	17.1	[7, 45]
Kaiho-2	210	5	215	2.3	[16]

The numbers of both sexes were combined. All the rats in Yururi-Moyururi were less than 9 months old. See ***Table 1.***

Table 2.
Percentage of the number of Norway rats that were 13 months old or more in Hahajima Islands, Yururi-Moyururi, Yokohama and Kaiho-2.

[32–34]. On Yururi-Moyururi, seven young rats less than 2 months old were pregnant in the summer (**Table 1**). Why did Norway rats on Yururi-Moyururi tend to mature at a young age and breed actively in the summer? It is possible that a protein-rich diet helped the rats to mature at a young age, as suggested by McCoy [5], who pointed out that a high-protein diet produces excellent reproductive conditions. Animal matter occupied 72.4 ± 39.8% (n = 38) by volume of the stomach contents of the rats in July–August 2013 in Yururi-Moyururi, and of this 11.9 ± 30.1% of rhinoceros auklets *Cerorhinca monocerata* Pallas [12]. From May to August, seabirds such as *Fratercula cirrhata* Pallas, *Cepphus carbo* Pallas, *Uria aalge* Pontoppidan, *Larus crassirostris* Vieillot, *Phalacrocorax urile* Gmelin, *P. capillatus* Temminck and Schlegel, and *P. pelagicus* Pallas also stay on Yururi-Moyururi to breed [35]. Norway rats probably prey on adults, nestlings, and eggs of these seabirds, which would supply the rats with sufficient nutrition to mature at a young age and engage in active breeding. Therefore, it is likely that Norway rats on Yururi-Moyururi depend on a diet of seabirds for their reproductive activities in the summer and a diet of carcasses of their own species under snow cover in the winter.

Norway rats preyed on adult *C. monocerata* irrespective of the body weight of the rats. The mean body weight of the predators, 187.7 ± 75.8 g (n = 16), was not significantly different ($P = 0.09$) from that of non-predators, 147.2 ± 54.2 g (n = 25) [36]. On the other hand, only larger roof rats on the Chichijima Islands in the Ogasawara Archipelago preyed on Bulwer's petrels *Bulweria bulwerii* Jardine and Selby, where the mean body weight of the predators, 201.6 ± 27.5 g (n = 22), was significantly larger ($P = 3.0 \times 10^{-4}$) than that of non-predators, 167.5 ± 35.4 g (n = 17) [36, 37]. Norway rats preyed on adults of *C. monocerata* (520 g [38]) that were larger than themselves, whereas roof rats preyed on adults of *B. bulwerii* (78–130 g [39]) that were smaller than themselves. These findings show that Norway rats are more aggressive predators of animal matter than roof rats [36].

As for the water supply for the rats, peat bogs are a source of water in Yururi but there are no peat bogs in Moyururi. However, the area around the Nemuro Peninsula is covered by dense sea fog for 101.4 days a year, and over 16 days per month between June and August [40]. Therefore, dew from dense sea fog is probably one of the water sources for Norway rats. I hypothesize that a process was established by which Norway rats have an appropriate diet and engage in water supply for survival and a bimodal cycle of reproduction in the summer and under the snow cover on Yururi-Moyururi.

3. Breeding in temperate climate zone

3.1 Breeding independent of season

Davis [41] reported that generally, the pregnancy rate in Norway rats is low in cold and hot seasons, and as a result, the rate shows a bimodal curve, with the highest peaks in the spring and autumn. The breeding season is usually estimated from pregnancy rates in adult females (percentages of visible pregnancies). However, recruitments of new generations in the population are more essential than pregnancy rates in population analysis [41, 42]. We can estimate the trend in the fluctuations of reproductive activities or recruitments based on age compositions even using surveys conducted once a year. Moors [43] discussed the age composition based on the age index estimated from the upper molars in Norway rat populations in Noises Island in New Zealand and concluded that recruitments were more active in the summer than in the winter. However, this age index revealed indefinite ages. Pucek and Lowe [44] recommended the eye-lens weight as the best criterion among the known indices for determining the age of small mammals. Then, Yabe et al. [45] analyzed age compositions based on the eye-lens criterion in Norway rat populations in February or March 2014–2016 in a 21-ha business district in Yokohama in the temperate climate zone (**Figure 1**). In this case, Norway rats showed recruitment peaks that were not always in the spring and autumn but also in the summer or winter, and the peaks changed every year (**Figure 3**). These results in Yokohama suggest that reproductive activities are controlled by factors other than temperature such as the food supply and environmental sanitation. In this business district in Yokohama, environmental sanitation activities conducted by volunteers control the Norway rat population [46].

3.2 Interruption of breeding by dehydration

Kaiho-2 (Fort No. 2) in Tokyo Bay (**Figure 1**; 4 ha, 35°18′ N, 139°44′ E) is an uninhabited islet in the temperate climate zone. This islet is covered with concrete, bricks, sand, sandy soil, grasses, herbs, and shrubs. Norway rats probably intruded into the islet in the early twentieth century, when a fort was constructed there. I discovered from the age compositions of the rats that their reproductive activities were interrupted around December or January [16, 17]. On average, between 1981 and 2010, in November, December, January, and February, the minimum temperatures were 9.6, 4.9, 2.3, and 2.6°C, and the amounts of precipitation were 107.0, 54.8, 58.9, and 67.5 mm, respectively, at Yokohama, a city close to Kaiho-2 [47]. Therefore, Kaiho-2 was dry in December and January compared with November and February. The water supply for Norway rats was probably insufficient around December and January because the amount of precipitation was low, the majority of succulent plants died, dew and standing water were limited, and the sandy soil lost moisture. Norway rats on the islet consumed protein-rich diets such as the mussel *Mytilus galloprovincialis* Lamarck and other marine invertebrates, which amounted to more than 50% of their stomach contents by volume, even in the winter [6]. However, such an invertebrate or protein-rich diet demands a large turnover of water [48]. Furthermore, most marine invertebrates including mussels are osmoconformers to the surrounding sea water [49]. Therefore, the interruption of reproductive activities during the winter was probably due to dehydration, but not to low temperature or food shortages.

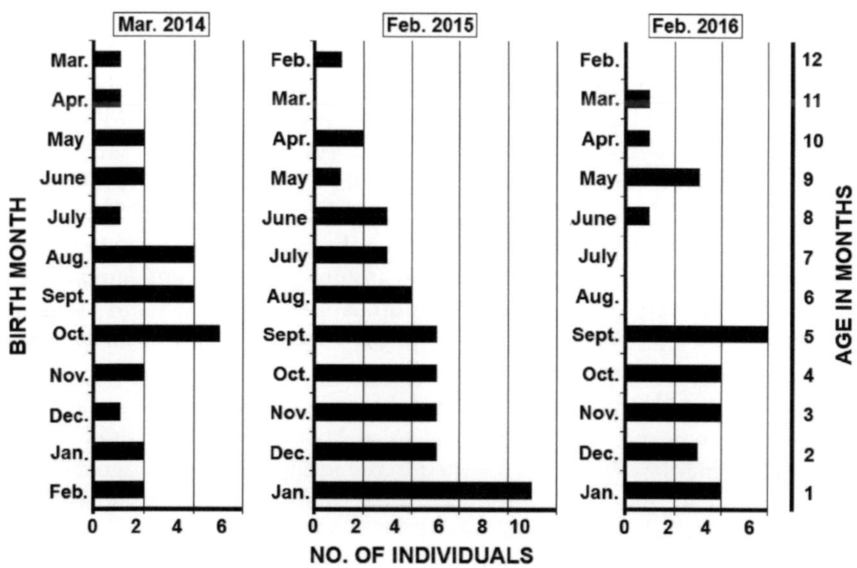

Figure 3.
Distributions of birth month and age in months in Norway rats caught in February or March 2014, 2015, and 2016 in Yokohama. Rats over 12 months old are excluded. Modified after [45].

		Body weight (g)	
Locality	Sex	3 Months	6 Months
Hahajima Islands	Male	77.0	112.6
	Female	79.5	102.9
Yururi-Moyururi	Male	208.3	278.5
	Female	156.3	205.1
Yokohama	Male	153.8	223.2
	Female	123.3	174.5
Kaiho-2 (non-wintered)	Male	193.3	281.3
	Female	168.9	259.9
Kaiho-2 (wintered)	Male	137.7	220.3
	Female	114.2	181.8

*Pregnant females are excluded. Body weights were calculated from regression lines. See **Figure 4**.*

Table 3.
Comparison of body weights of 3- and 6-month-old Norway rats living in the Hahajima Islands, Yururi-Moyururi, Yokohama, and Kaiho-2 (non-wintered and wintered groups) [7].

	Wintered*	Non-wintered*	Pregnant females
Kaiho-2	0.10 ± 0.04^a	0.16 ± 0.06^b	0.22 ± 0.06^c
n =	28	63	9
Shikine-jima	0.11 ± 0.06^a	0.12 ± 0.06^a	0.19 ± 0.07^c
n =	37	40	4

**Excluding pregnant females.*
Fat indexes were significantly different (t-test, $p < 0.05$) if they are followed by different letters. FI = 1.01FI' + 0.01, where FI' = fat free dry weight/dry weight [51].

Table 4.
Comparison of fat index (FI, mean ± SD) between wintered and non-wintered Norway rats on Kaiho-2 and a forested island (Shikine-jima) [17].

Collier and Levitsky [50] showed that albino *R. norvegicus* rats lose their body mass to maintain water balance when the water supply is insufficient. Moors [43] suggested that shortages of protein-rich diets and fresh water restrict the sexual maturity of females, litter sizes, and the growth of juveniles in Norway rats on Noises Island in New Zealand. It is likely that a similar situation occurred in Norway rats on Kaiho-2. The age composition of Norway rats on this islet showed a gap between the generations borne before and after the season around December and January, when breeding was interrupted. As a result, their population was divided into a wintered group and a non-wintered group based on the gap. The body mass of the wintered group was lower than that of the non-wintered group (**Table 3**). I compared a body fat index determined by the method of Yabe [51] among the wintered group, the non-wintered group, and pregnant females. Also, I compared the index between Kaih-2 and Shikine-jima (a 390-ha forested island in the Izu Archipelago, 34°19′ N, 139°12′ E) (**Table 4**). As a result, I found that the small body mass in the wintered group in Kaiho-2 was due to body fat loss [17]. The body fat indexes showed that pregnant females kept a high level of body fat irrespective of whether they were in the wintered or non-wintered group, or on Kaiho-2 or Shikine-jima. Pregnant females deposit body fat for reproduction, probably because they require more energy than nonreproducing females as was pointed out by Robbins [52]. The lost body fat in the wintered group was not recovered after the dehydration period, and the non-wintered group kept a high level of body fat [16, 17]. This fat deposition procedure is different from that in mammals, which deposit body fat as a prelude to times when the energy intake will be less than the energy expenditure [52].

4. Dehydration and low body mass in subtropics

The Ogasawara Archipelago (Bonin Islands, Ogasawara Islands) is composed of the Mukojima Islands, the Chichijima Islands, the Hahajima Islands, and the Kazan (Volcano) Islands in the subtropics (**Figure 1**). Norway rats are thought to have intruded into the Ogasawara Archipelago between 1660 and 1862, but now they are living only in the Hahajima Islands and the Kazan Islands [53–55]. On the other hand, roof rats are prosperous and are distributed in most islands in the archipelago [56, 57], although they intruded there in the 1910s or 1920s, later than the Norway rats [54, 58]. It remains to be clarified why Norway rats are restricted to only a few islands in the archipelago.

The body mass of Norway rats on the Hahajima Islands is about half the weight of Norway rats on Yururi-Moyururi, Yokohama, and Kaiho-2 (**Table 3** and **Figure 4**). The low mass of the Hahajima rats was due to environmental factors rather than genetic factors such as Bergman's rule and the founder's effect. This was proved by the fact that the head and body length, tail length, and length of the upper molar row were not significantly different between the rats from Hahajima and those from other localities [7]. Therefore, the skeletons were the same but the body masses were different between the Hahajima rats and the others.

Norway rats on the Hahajima Islands tended to feed on plant matter such as fruits and seeds (95.2 ± 21.8%, n = 21, by volume percentage in stomach contents) and no seashore animals were found even in rats living close to the seashore [7]. This is an abnormal food habit in the Norway rat, which prefers animal matter [6]. As I previously mentioned, preying on plant matter helps maintain water balance because the consumption of animal matter or of a protein-rich diet requires more water intake. However, this change in food habits may lead to a protein deficiency and body weight loss in the rats. To meet their energy requirements, mammals consume their gastrointestinal contents first, but finally they utilize their body fat and protein, which leads to long-term weight loss [52]. Moors [43] suggests that a shortage of protein-rich diets and fresh water limited the reproductive activities of Norway rats on Noises Island in New Zealand. It is likely that on the Hahajima Islands as well, protein deficiency and dehydration decrease the weight and inactivated the reproduction of Norway rats. I suppose that Norway rats on the Hahajima Islands are less aggressive predators than rats living in the other habitats because of their food habit.

The Ogasawara Archipelago is probably an uncomfortable habitat for Norway rats due to chronic dehydration, which restricts their distribution. In the Hahajima Islands, there are streams and ponds on the main island but not on the surrounding islands. However, Norway rats were found even on the surrounding islands and in areas far from such water sources [7]. Therefore, dehydration in Norway rats on the Hahajima Islands was not due to a lack of such water sources. The mean annual precipitation in the Chichijima Islands from 1971 to 2000 was 1280 mm, and the mean potential evaporation (the amount of evaporation that would occur when enough water is given) was 1380 mm [27]. The former is less than the latter, and as a result, the soil tends to be dry. This indicates a potential cause of dehydration in Norway rats. However, the Hahajima Islands, with a mountain 462 m in height, is foggy and more humid than the Chichijima Islands, with a mountain 326 m in height, and the

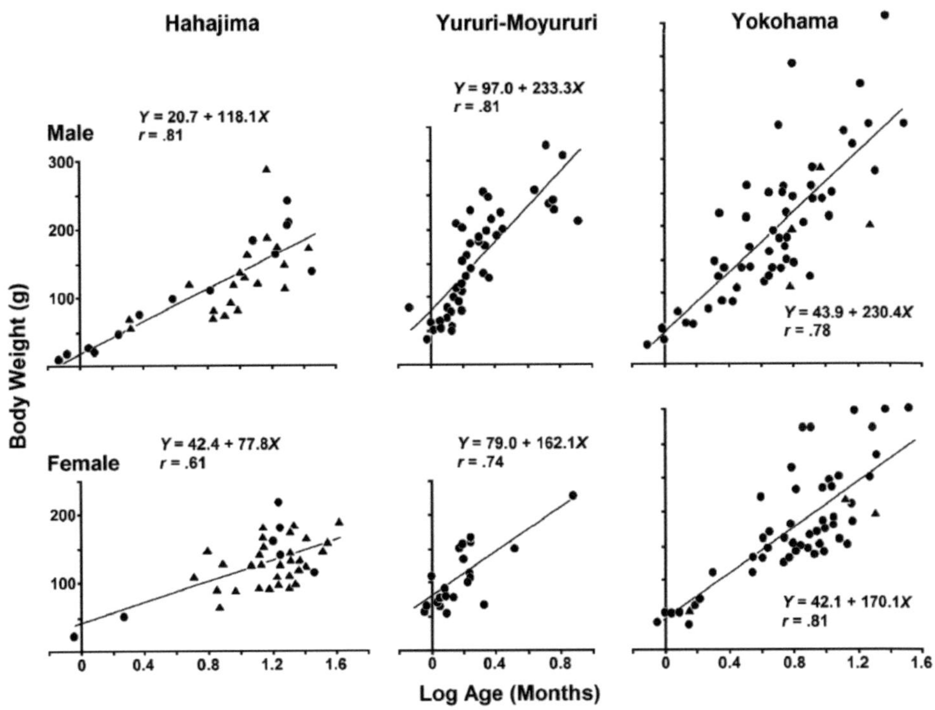

Figure 4.
*Comparison of body weight in grams (Y) and log value of age in months (X) and resulting regression lines for male and female Norway rats from the Hahajima Islands, Yururi-Moyururi, and a business district in Yokohama excluding pregnant females, showing infection with rat lungworms (*Angiostrongylus cantonensis*) [7]. Circles and triangles show rats that were negative and those that were positive for the infection, respectively.*

low and flat Mukojima Islands [27]. Therefore, Norway rats probably thrive better in the Hahajima Islands than in the others.

Renal structures show that the ability to concentrate urinary water in Norway rats is like that in roof rats [59]. However, protein-rich diets demand a larger turnover of water than diets rich in carbohydrates or fat [48], and Norway rats feed on protein-rich diets, whereas roof rats prefer plant matter to animal matter [6]. Therefore, Norway rats require more water intake than roof rats. This difference in water requirements is probably one of the factors that separate the two species in the geographical distribution especially in tropical and subtropical climate zones [15]. Mild temperature is a secondary factor in determining the Norway rat distribution, after water balance and an appropriate diet. Even in the tropical climate zone, Norway rats are prosperous in large cities such as Bangkok (13° 44′ N, 100° 29′ E) and Chanthaburi (12° 36′ N, 102° 06′ E) in Thailand, which are surrounded by networks of watercourses and damp environments [15, 60]. Generally, in these habitats, there are protein-rich diets including garbage and invertebrates such as earthworms and insects [6]. Therefore, protein-rich diets and the means for avoiding dehydration such as creeks and sewage provide Norway rats with thriving habitats in large cities. These facts suggest that diets rich in animal matter or protein are associated with water balance, which are essential factors in the geographical distribution of Norway rats.

5. Conclusion

Mild temperature is a secondary factor in the reproductive activities of Norway rats as was proved by the results in Yururi-Moyururi in the subarctic zone and in an urban area in Yokohama in the temperate zone. In Yururi-Moyururi, the rats recruited new generations in their population under snow cover probably by preying on remnants of their own species, which were left by birds of prey such as common buzzards. In Yokohama, the rats showed peaks of recruitment even in the summer and winter, though the season of the peaks changed every year. Even in the tropics, the rats are prosperous in large cities such as Bangkok and Chanthaburi in Thailand, which are surrounded by networks of watercourses and damp environments [15, 60]. It is likely that watercourses supply the rats with an appropriate diet discarded from houses as well as with moist conditions.

Water balance and a protein-rich diet are essential factors in the reproductive activities and distribution of Norway rats as was shown by the results in Kaiho-2 and the Hahajima Islands. The rats on Kaiho-2 in the temperate zone stopped recruiting of new generations and lost body mass by consuming their body fat in the winter because of dehydration. In the Hahajima Islands in the subtropics, the rats fed mainly on plant matter to maintain water balance because of chronic dehydration, and as a result, they lost body mass. In this case, the rats probably avoided consuming animal matter or a protein-rich diet to maintain water balance, but they consumed protein from within their bodies instead. Norway rats usually feed on a protein-rich diet or animal matter, which differs from the food habits of roof rats, which prefer plant matter to animal matter (6). Thus, a protein-rich or animal matter diet is an appropriate diet for Norway rats.

Author details

Tatsuo Yabe
Rat Control Consulting, Yamato, Japan

*Address all correspondence to: rccty@js8.so-net.ne.jp

References

[1] Atkinson IAE. The spread of commensal species of *Rattus* to oceanic islands and their effects on island avifaunas. In: Moors PJ, editor. Conservation of Island Birds: Case Studies for the Management of Threatened Island Species. ICBP Technical Publication No. 3. Cambridge: International Council for Bird Preservation; 1985. pp. 35-81

[2] Brooks JE, Rowe FP. Commensal Rodent Control. Geneva: World Health Organization; 1987. p. 107

[3] Twigg G. The Brown Rat. London: David & Charles; 1975. p. 150

[4] Tomich PQ. Mammals in Hawaii. 2nd ed. Hawaii: Bishop Museum Press; 1986. p. 375

[5] McCoy RH. Dietary requirements of the rat. In: Farris EJ, Griffith JQ Jr, editors. The Rat in Laboratory Investigation. 2nd ed. Philadelphia: JB Lippincott; 1949. pp. 68-103

[6] Yabe T. The relation of food habits to the ecological distributions of the Norway rat (*Rattus norvegicus*) and the roof rat (*R. rattus*). Japanese Journal of Ecology. 1979;**29**:235-244

[7] Yabe T, Horikoshi K, Hashimoto T. Small mass of *Rattus norvegicus* (Rodentia: Muridae) on the Ogasawara Islands, Japan. Pacific Science. 2017;**71**:257-268. DOI: 10.2984/71.3.2

[8] Moors PJ, Atkinson IAE. Predation on seabirds by introduced animals, and factors affecting its severity. In: Croxall JP, Evans PGH, Schreiber RW, editors. Conservation of Island Birds. ICBP Technical Publication No. 2. Cambridge: International Council for Bird Preservation; 1984. pp. 667-690

[9] Paul E. The Rat Island Eradication Project: A Critical Evaluation of Nontarget Mortality. Ornithological Council: Maryland; 2010. p 85

[10] Pye T, Bonner N. Feral brown rats, *Rattus norvegicus*, in South Georgia (South Atlantic Ocean). Journal of Zoology (London). 1980;**192**:237-255

[11] Schiller EL. Ecology and health of *Rattus* at Nome, Alaska. Journal of Mammalogy. 1956;**37**:181-188

[12] Yabe T, Minato R, Hashimoto T. Breeding under snow cover in Norway rats (*Rattus norvegicus*) on uninhabited islands in Hokkaido, Japan. Russian Journal of Theriology. 2017;**16**:43-46. DOI: 10.15298/rusjtheriol.16.1.04

[13] Corbet GB, Hill JE. The Mammals of the Indomalayan Region: A Systematic Review. Oxford: Oxford University Publications; 1992. p. 488

[14] Walton DW, Brooks JE, Tun UMM, Naing UH. The status of *Rattus norvegicus* in Rangoon, Burma. Japanese Journal of Sanitary Zoology. 1977;**28**:363-366

[15] Yabe T, Chenchittikul M. Predominant species of commensal rats in rural areas in Thailand. Japanese Journal of Sanitary Zoology. 1985;**40**:345-347

[16] Yabe T. Influence of dehydration on breeding, body size and kidney structure in Norway rats (*Rattus norvegicus*) on an islet. Physiology and Ecology Japan. 1982;**19**:7-13

[17] Yabe T. Fat deposits for wintering in the Norway rat, *Rattus norvegicus*. Journal of Mammalogical Society of Japan. 1994;**19**:129-133

[18] Yabe T. Protein-rich diet and water balance as limiting factors in the geographical distribution of the Norway rat, *Rattus norvegicus* (In Japanese with English abstract). Medical Entomology and Zoology. 2018;**69**:41-47. DOI: 10.7601/mez.69.41

[19] Yabe T. Eye lens weight as an age indicator in the Norway rat. Journal of Mammalogical Society of Japan. 1979;**8**:54-55

[20] Russell JC, Towns DR, Clout MN. Review of Rat Invasion Biology: Implications for Island Biosecurity. Science for Conservation 286. Department of Conservation: Wellington; 2008. p. 53

[21] Kucera E, Fuller WA. A winter study of small rodents in aspen parkland. Journal of Mammalogy. 1978;**59**:200-204

[22] Inukai T. Outbreak of Norway rats and the damage caused by them in Sakhalin (In Japanese). Plants and Animals. 1939;7(12):57-69

[23] Whitney P, Feist D. Abundance and survival of *Clethrionomys rutilus* in relation to snow cover in a forested habitat near college. In: Merritt JF, editor. Winter Ecology of Small Mammals. Pittsburgh: Carnegie Museum of Natural History; 1984. pp. 113-119

[24] Maeda M. Flowing, fruiting and withering of the bamboo grass *Sasa kurilensis*, and rodents (In Japanese). Sapporo Rin-Yu. 1977;**188**:43-54

[25] Inukai T. Outbreak of Norway rats in Sakhalin and review of the control measures (In Japanese). Journal of Agriculture and Forestry Society of Sapporo. 1942;**34**(3):1-13

[26] Kinoshita E. Field Rodents and their Control Method (In Japanese). Sapporo: Hokkaido Shinrin-Boeki Kyokai; 1965. p. 60

[27] Kanto Regional Environmental Office, Tokyo Metropolitan Government, Ogasawara Village. Ogasawara Islands Ecosystem Conservation Action Plan [Internet]. 2010. Available from: ogasawara-info.jp/pdf/isan/ActionPlan _eigo.pdf [Accessed: 14 February 2018]

[28] Wild Bird Society of Japan. Important Bird Area in Japan—Yururi-Moyururi [Internet]. 2006. Available from: www.wbsj.org/nature/hogo/others/iba/search/sites/hokkaido/14-yururi.htm [Accessed: 29 July 2016]

[29] Errington PL. Wintering of field-living Norway rats in south-central Wisconsin. Ecology. 1935;**16**:122-123

[30] Selås V, Tveiten R, Aanonsen OM. Diet of common buzzards (*Buteo buteo*) in southern Norway determined from prey remains and video recordings. Ornis Fennica. 2007;**84**:97-104

[31] Meehan AP. Rats and Mice, their Biology and Control. East Grinstead: Rentokil; 1984. p. 183

[32] Farris DJ. Breeding of rat. In: Farris EJ, Griffith JQ Jr, editors. The Rat in Laboratory Investigation. 2nd ed. Philadelphia: JB Lippincott; 1949. pp. 1-18

[33] Pascal M, Lorvelec O. *Rattus norvegicus*. DAISIE Fact Sheet [Internet]. 2006. Available from: www.europe-aliens.org/pdf/Rattus_ norvegicus.pdf [Accessed: 30 May 2018]

[34] Quesenberry KE, Boschert KR. Breeding Reproduction of Rats. Merck Veterinary Manual [Internet]. 2018. Available from: https://www.merckvetmanual.com/all-other-pets/rats/breeding-and-reproduction-of-rats [Accessed: 30 July 2018]

[35] Kushiro Environmental Office of Japan. Report on a Trial to Conserve the National Sanctuary of Yururi-Moyururi. [Internet]. 2017. p. 12 Available from: http://hokkaido.env.go.jp/kushiro/161116w1.pdf [Accessed: 25 February 2018]

[36] Yabe T. Behavioral differences in predation on seabirds between roof

rats and Norway rats (In Japanese). Annual Report of Ogasawara Research. 2019;**42**:23-30

[37] Yabe T, Hashimoto T, Takiguchi M, Aoki M, Kawakami K. Seabirds in the stomach contents of black rats *Rattus rattus* on Higashijima, the Ogasawara (Bonin) Islands, Japan. Marine Ornithology. 2009;**37**:285-287

[38] Brazil M. Field Guide to the Birds of East Asia. London: Christopher Helm; 2009. p. 528

[39] del Hoyo J, Elliot A, Sargatal J. Handbook of the Birds of the World. Vol. 1. Barcelona: Lynx Editions; 1992

[40] Sapporo Meteorological Observatory. Climate in the Pacific Region of Hokkaido [Internet]. 2018. Available from: www.jma.go.jp/sapporo/tenki/kikou/tokucho/column_fog [Accessed: 14 February 2018]

[41] Davis DE. The characteristics of rat populations. Quarterly Review of Biology. 1953;**28**:373-401

[42] Davis DE, Golley FB. Principles in Mammalogy. New York: Van Nostrand Reinhold; 1965. p. 335

[43] Moors PJ. Norway rats (*Rattus norvegicus*) on the Noises and Motukawao Islands, Hauraki Gulf, New Zealand. New Zealand Journal of Ecology. 1985;**8**:37-54

[44] Pucek Z, Lowe VP. Age criteria in small mammals. In: Golley FB, Petrsewicz K, Ryszkowski L, editors. Small Mammals: Their Productivity and Population Dynamics. Cambridge: Cambridge University Press; 1975. pp. 55-72

[45] Yabe T, Otomo T, Harashima T, Shigeoka H, Yamaguchi K. Breeding season in urban population of Norway rats in Yokohama estimated from age composition (In Japanese with English abstract). Medical Entomology and Zoology. 2016;**67**:199-202. DOI: 10.7601/mez.67.199

[46] Yabe T, Otomo T, Harashima T, Shigeoka H, Yamaguchi K. Rat control campaign by Kanagawa Pest Control Association in urban Yokohama (In Japanese with English abstract). Pestology. 2012;**27**:7-11

[47] Meteorological Agency. Past Meteorological Data [Internet]. 2018. Available from: www.data.jma.go.jp/gmd/risk/obsde [Accessed: 16 February 2018]

[48] Fitzsimons TJ, Le Magnen J. Eating as a regulatory control of drinking in the rat. Journal of Comparative Physiology and Psychology. 1969;**67**:273-283

[49] Schmidt-Nielsen K. Animal Physiology. Adaptation and Environment. London: Cambridge University Press; 1975. p. 699

[50] Collier G, Levitsky D. Defense of water balance in rats: Behavioral and physiological responses to depletion. Journal of Comparative Physiology and Psychology. 1967;**64**:59-67

[51] Yabe T. A simple method for determining fat deposits in rodents. Journal of Mammalogical Society of Japan. 1992;**16**:97-100

[52] Robbins CT. Wildlife Feeding and Nutrition. 2nd ed. London: Academic Press; 1993. p 352

[53] Yabe T. “Mizu-Nezumi (the water rat)” on Ogasawara (In Japanese). Annual Report of Ogasawara Research. 2006;**29**:19-22

[54] Yabe T. Control of Commensal Rodents in Japan (In Japanese). Tokyo: Chijinshokan; 2008. p. 169

[55] Yabe T, Matsumoto T. A survey on the murine rodents on Chichijima and Hahajima, the Ogasawara Islands.

Journal of Mammalogical Society of Japan. 1982;**9**:14-19

[56] Horikoshi K, Suzuki H, Sasaki T, Chiba H. The impact assessment of invasive alien mammals on sea bird colonies (In Japanese). Global Environmental Research. 2009;**14**:103-105

[57] Kawakami K. Feral cats and rodents in the Bonin Islands (In Japanese). In: Ecological Society of Japan, editor. Handbook of Alien Species in Japan. Tokyo: Chijinshokan; 2002. pp. 236-237

[58] Kuroda C. Mammals of the Ogasawara Islands (In Japanese). Bulletin of the Biogeographical Society of Japan. 1930;**1**:81-88

[59] Yabe T. Renal structural indices for the ability to conserve water in rodents, *Mus molossinus*, *Rattus norvegicus*, and *R. rattus*. Physiology and Ecology Japan. 1983;**20**:53-57

[60] Chenchittikul M, Daengpium S, Hasegawa M, Itoh T, Phanthumachinda B. A study of commensal rodents and shrews with reference to parasites of medical importance in Chanthaburi province, Thailand. Southeast Asian Journal of Tropical Medicine and Public Health. 1983;**14**:255-259

Chapter 4

Nesting Behavior of Indian Giant Squirrel (*Ratufa indica* Erxleben, 1777) in Mudumalai Tiger Reserve, Western Ghats, Southern India

Samson Arockianathan

Abstract

The present study was carried out on the nesting behavior of IGS in the Mudumalai Tiger Reserve during the month of June 2015 to June 2017 (2 years). A total of 192 nesting trees with 279 nests belong to 19 tree species were identified as nesting trees preferences of IGS. Of which *Bambusa arundinacea* grass species was the dominant nesting grass species of the IGS in Mudumalai Tiger Reserve (11%, n = 22). The overall nest height of the IGS was 19.70 m and a maximum height of 34 m and a minimum height of 8 m. The nest direction shows that the North East held the number of the nest (n = 137), and the nest position shows that the Crown (n = 197) contained the number of the nest. The nest position shows that top (n = 220) were contained the number of nests compared to the middle (n = 59). On the other hand, no nest was placed on the down position.

Keywords: Indian giant squirrel, Mudumalai Tiger Reserve, nesting, Western Ghats

1. Introduction

The Indian or Malabar giant squirrel (*Ratufa indica* Erxleben, 1777) is endemic to Peninsular India (South India) [1]. Although it is widely distributed within its range, it occurs in severely fragmented populations [2]. It has faced local extinction and range restriction in several areas due to hunting and habitat loss and suitable habitat is limited in the areas where it occurs [3]. The Indian giant squirrel is currently listed in the “Least Concern” category of IUCN Red List, Appendix II of CITES and Schedule II of the Wildlife (Protection) Act, 1972 of India [3, 4]. The Indian giant squirrel occurs in the elevation range of 180–2300 m and inhabits deciduous, mixed deciduous and moist evergreen forests [5]. It is a large-bodied (90–100 cm), diurnal and arboreal squirrel [6]. A solitary living species, it is seen in pairs only during the breeding season. It usually constructs more than one nest, or drey, within a single breeding season. The nests, which are made of leaves and twigs, are built-in tall, profusely branched trees, in the higher canopy [7, 8]. The species is omnivorous and feeds on fruits, flowers, nuts, bark, bird eggs and insects [8, 9]. The ecology of squirrels from Asian countries has been little studied and published information is scarce [10, 11]. In Mudumalai Tiger Reserve, there is an only one literature was available on IGS population and nesting ecology [12]. Hence the present study was under took major objectives on (1) To find out the nesting tree preference, (2) To find out the nesting trees variables to support the IGS nesting, (3) Nesting behavior of IGS, and (4) To given an scientific recommendation for long term management and sustainable conservation of the species.

2. Study area

Mudumalai Tiger Reserve is one of the few areas in the country with a rich and varied terrain, flora and fauna. Mudumalai plays an important role in biodiversity conservation of especially large mammals, by offering habitat contiguity of about 3300 km^2 with three other protected areas in the region, namely Nagarahole and Bandipur National Park and Wayanad Wildlife Sanctuary through forest corridors between the Western Ghats and the Eastern Ghats. The reserve was created in 1940, the first in southern India, with an area of 60 km^2. In 1956, it was enlarged to 295 km^2 and later to a further 321 km^2 and 688.59km^2 core zone = 321 km^2 and buffer zone = 367.59 km^2 which it is present extent (**Figure 1**). Champion and Seth [13] classified the vegetation type in Mudumalai as Southern Tropical dry thorn forest, Southern Tropical dry deciduous forest, Southern Tropical moist deciduous forest, Southern Tropical semi-evergreen, Moist bamboo brakes, and Riparian fringing forest.

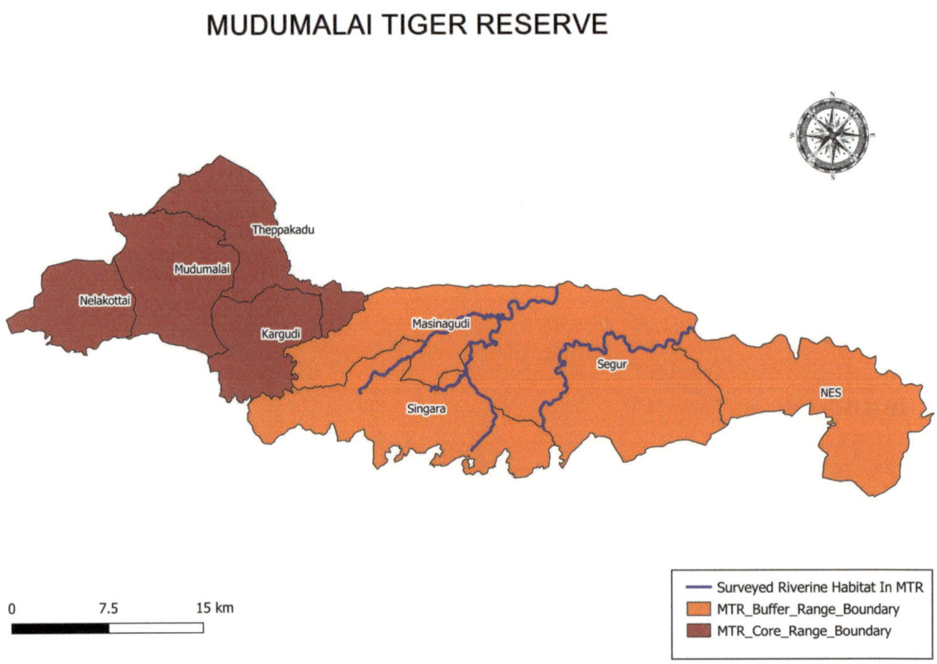

Figure 1.
Map showing the surveyed riverine habitats in Mudumalai Tiger Reserve.

3. Methodology

3.1 Data collection

Data were collected from June 2015 to June 2017 mostly on breeding seasons when the squirrels are more active and easily seen. We searched for animals and their nests along the natural trails in the dry thorn forest. Most of the nesting trees were located through intensive searches in the area by inspecting potential nesting trees and nests. The presence of IGS and their activity provide indirect evidence of use as nest trees. The IGS nesting trees were marked with GPS coordinates and classified with identification. The quantification of nesting habitat followed methods suggested by James and Shugart [14] and subsequently by Kannan [15], Mudappa and Kannan [16], and Girikaran et al. [17]. Vegetation and nest tree parameter was quantified in circular plots of 15 m (0.07 ha) with the nest tree as the center. All the trees (GBH > 25 cm) were enumerated and GBH (Girth at Brest Height) measured. Canopy cover was visually estimated. The elevation of the nesting tree distances to the nearest road, habitation was also noted. The nest tree parameters were measured such as tree height, basal area, diameter at breast height, number of primary branches and secondary, canopy cover, canopy height, canopy width and tree status such as (dead or alive) were noted. Such parameters were also quantified in similar-sized plots located 100 m in a random direction from the nest tree, where the nearest tree of GBH > 250 cm was chosen as the centre tree and the same nest tree parameters were also taken into the account for comparison of random (non-nest) plots with nest tree plots were made to determine parameters likely to affects choice of nesting habitat by Indian Giant Squirrel. The availability and density of potential nest tree species were assessed from 16 0.25 ha (50 m × 50 m) vegetation plots (2.5 ha).

3.2 Statistical treatment

Mean (M) and Standard Error (SE) was calculated to the nesting trees variables in the study area. Pearson's correlation coefficient matrix was performed to understand the variables significances among the nesting trees. Man Whitney U test was used to determine differences in 13 parameters between nest (n = 158) and non-nest (n = 250) plots. Principal Component Analysis was used to understand nest site selection. Statistical analyses were performed using *Graph Pad Prism 5 and SPSS* 17.0 statistical computer software.

4. Result

A total of 192 nesting trees with 279 nests belonging to 19 trees species were identified as nesting trees preferences of IGS in the Mudumalai Tiger Reserve (**Table 1**). Of which *Bambusa arundinacea* was the dominant nesting grass species of the IGS (11%, n = 22) followed by *Terminalia arjuna* (10%, n = 20), *Spondias mangifera* (9%, n = 18), *Syzygium cumini* (7%, n = 14) and *Ficus benghalensis*, *Manilkara hexandra*, *Sapindus emarginata* were each 12 nesting trees (n = 6%), respectively. Among the nest, wise number of nests were recorded in the *Bambusa arundinacea* (20%, n = 56) followed by *Terminalia arjuna* (10%, n = 28), *Spondias mangifera* (9%, n = 26), *Syzygium cumini* (8%, n = 22) and *Ficus benghalensis* (6%, n = 16). There is a significant difference were observed on nesting trees preferences as well as the number of nests in a nesting tree (t = 2.539; P = 0.0184). The overall nest height of the IGS was 19.70 ± 3.25 m and a maximum height of 34 m and a minimum height of 8 m and the nest direction shows that North East has held the number of nests (n = 137) followed by South East (n = 83), South West (n = 40) and North West (n = 19) (**Figure 2**). The nest position shows that Crown (n = 197) were contained the number of nest camper to lumb (n = 82). The nest position shows that top (n = 220) were contained the number of nests compare to the middle (n = 59). on the other hand, no nest was placed on the down position. A total of 14 variables

S.No	Scientific name of the nesting trees	Number of nesting trees	Relative abundances of the nesting trees %	Number of nests	Relative Abundances of the nest in nesting trees %
1	*Bambusa arundinacea* (Grass species)	22	11	56	20
2	*Terminalia arjuna*	20	10	28	10
3	*Spondias mangifera*	18	9	26	9
4	*Syzygium cumini*	14	7	22	8
5	*Ficus benghalensis*	12	6	16	6
6	*Manilkara hexandra*	12	6	14	5
7	*Sapindus emarginata*	12	6	12	4
8	*Ailanthus excelsa*	10	5	12	4
9	*Terminalia bellirica*	10	5	10	4
10	*Acasia leucophloea*	8	4	8	3
11	*Schleichera oleosa*	8	4	8	3
12	*Tamarindus indica*	8	4	14	5
13	*Albizia lebbeck*	7	4	7	3
14	*Terminalia crenulata*	7	4	10	4
15	*Cassine glauca*	6	3	12	4
16	*Pongamia pinnata*	6	3	8	3
17	*Ficus mollis*	5	3	7	3
18	*Ficus microcarpa*	4	2	6	2
19	*Filicium decipiens*	3	2	3	1
	Total	192		279	

Table 1.
Nesting tree preference of IGS in the Mudumalai Tiger Reserve.

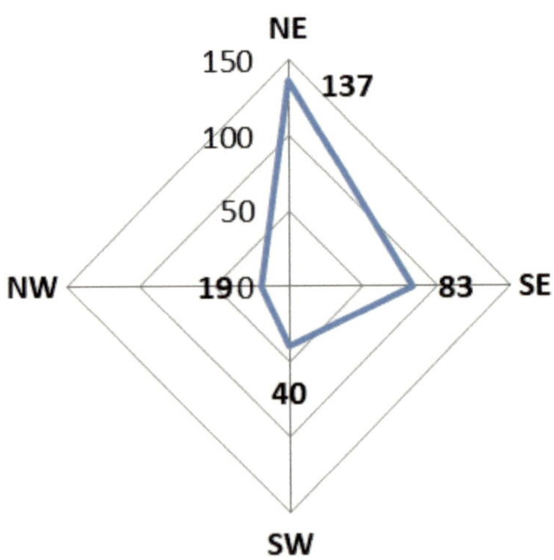

Figure 2.
Nest direction in nesting trees in Mudumalai Tiger Reserve.

were collected from the nesting trees and non-nesting trees for analyzing the preferences of IGS nesting in the study area (**Table 2**).

Fourteen variables of nest tree and centre tree of non-nest sites were measured and are given above (**Table 2**). Nest trees differed significantly from centre trees of non-nest plots, in terms of size. The height of the tree, basal area, GBH, Branch end, secondary branches, canopy length, canopy cover, and elevation were all significantly greater in nest trees than non-nest centre trees (**Table 2**). But there was no significant difference in, Branch start, branch start, and branch end distance, Primary branches, canopy width, Distance to human habitation and distances to the road between nest plot and centre trees of non-nest plots. However, there was a significant difference in large tree density (GBH ≥ 25 cm, GBH ≥ 26–75 cm, GBH ≥ 126–175 cm and GBH ≥ 326–375 cm) between the nest and non-nest plots (**Table 2**).

Variables	Nest plot (n = 192)	Non-nest plot (n = 250)	*U*	*P < 0.05*
Nest/centre tree height (m)	25.63 ± 0.68	21.48 ± 0.47	2750	0.40*
Nest/centre tree basal Area (cm)	423.56 ± 19.28	389.35 ± 14.78	2125	0.03*
Nest/centre tree girth at breast height (cm)	397.32 ± 17.25	358.46 ± 16.25	2091	0.00*
Nest/centre tree branch start (m)	7.35 ± 0.29	5.46 ± 0.37	2093	0.12
Nest/centre tree branch end (m)	23.35 ± 0.46	19.27 ± 0.49	2782	0.05*
Nest/centre tree branch start and branch end distance (m)	9.18 ± 0.37	10.23 ± 0.29	3272	0.77
Nest/centre tree primary branches	3.68 ± 0.19	5.16 ± 0.13	2951	0.17
Nest/centre tree Secondary branches	42.37 ± 1.27	44.19 ± 0.78	2901	0.03*
Nest/Centre tree Canopy length (m)	27.53 ± 0.68	28.32 ± 0.83	2466	0.00*
Nest/Centre tree canopy width (m)	28.34.88 ± 0.75	29.56 ± 0.43	2992	0.22
Nest/Centre tree canopy cover (%)	78.32 ± 2.35	82.34 ± 1.36	3078	0.05*
Nest/centre tree Distance to human habitation (km)	10.36 ± 1.26	11.12 ± 0.18	3298	0.54
Nest/centre tree Distance to road (km)	9.09 ± 0.35	8.63 ± 0.52	3259	0.37
Nest/centre tree elevation (m)	893.68 ± 19.84	887.37 ± 17.39	4235	0.02*
Tree density/ha				
i. Trees of GBH ≥ 25 cm	28.35 ± 0.54	26.53 ± 0.69	2833	0.05*
ii. Trees of GBH ≥ 26–75 cm	5.17 ± 0.16	4.52 ± 0.32	2496	0.00*
iii. Trees of GBH ≥ 76–125 cm	3.36 ± 0.17	3.85 ± 0.19	3612	0.35
iv. Trees of GBH ≥ 126- 175 cm	3.13 ± 0.29	2.98 ± 0.93	2843	0.04*
v. Trees of GBH ≥ 176–225 cm	2.38 ± 0.59	1.87 ± 0.53	3036	0.16
vi. Trees of GBH ≥ 226–275 cm	1.25 ± 0.17	1.89 ± 0.23	3314	0.65
vii. Trees of GBH ≥ 276- 325 cm	1.35 ± 0.29	1.59 ± 0.12	3194	0.37
viii. Trees of GBH ≥ 325 cm	1.13 ± 0.75	1.38 ± 0.74	2841	0.03*
ix. Trees of GBH ≥ 376–425+ cm	0.52 ± 0.46	0.75 ± 0.49	3217	0.04*

**Parameters that was significantly different between nest and non-nest plot P < 0.05 significant.*

Table 2.
Characteristics of nest-site and non-nest site plots of IGS in the Mudumalai Tiger Reserve.

The principal component analysis (PCA) was carried out using the nest site characteristics data from all the nests of IGS observed (n = 158). **Table 3** shows Pearson's correlation matrix between the 14 variables.

PCA extracted three principal components that elucidated 87.12% variability (**Table 4**). The first component explained 39.02% variability that gives details of seven nest tree variables such as tree height, Branch Start, Branch End, Branch Start and Branch End Distance, Canopy Width, Distance to human habitation and Distance to Road in the plot and that were positively correlated with the first component. High values on the first component corresponding to the tallness of nest trees, Branch Start, Branch End, Branch Start and Branch End Distance, Canopy Width. Thus, the first component represents, with increasing values, the size of the nest tree and tallness will also increase. The first component was also positively correlated to Distance to human habitation and Distance to Road variable, which indicates, with increasing values, greater distance to human habitation and roads. The second component explained 29.07% variability that explained five nest tree variables such as basal area, GBH, Primary Branch, Secondary Branch and Canopy Length (**Table 4**). High values on the second component correspond to a basal area GBH, Primary Branch, Secondary Branch and Canopy Length. Thus, the second component also represents, with increasing values, the size of the nest tree and basal area and branch structure of the tree will also increase. The third component explained 11.68% of the total variance and was related to Canopy cover and human habitation. The fourth component explains 7.35% of the total variance and was related to Canopy cover and Elevation (**Table 4**).

A total of 24 potential nest tree species of IGS that occurred at the study area was identified based tree genera or species those that generally attain a large tree size (**Table 5**). Of these only 19 species were used for nesting by IGS in the Mudumalai Tiger Reserve. All of these trees were emergent, large girth trees and are relatively more common than other species; in fact, *Terminalia arjuna* was the most common tree species among these. The overall occurrence of *Terminalia arjuna* was 29.3 trees per ha and 10.4 per ha for trees of GBH $\geq$ 250 cm which were recorded in the 146 0.25 ha plots during the study period (**Table 5**). Density of large trees (GBH $\geq$ 250 cm) species *Alianthus excelsa* 8.91 per ha recorded in the 146 0.25 ha plots and *Pongamia pinnata* 5.89 per ha, *Manilkara hexandra* 4.40 per ha, *Schleichera oleosa* 3.46 per ha and *Spondias mangifera* 2.46 per ha recorded in the 146 0.25 ha plots. According to the tree size, the overall availability of the species was an important factor in the nest tree selection by Indian Giant Squirrel. In the study plots covering 36.5 ha, the overall availability of trees GBH $\geq$ 250 cm was 14.75 per ha (56.56 trees) (**Table 5**).

5. Discussion

Preference for nesting trees could depend on factors such as access to nesting material and food, nest safety and the branching pattern of the tree species. A total of 18 tree species and one grass species were recognized as nesting trees of IGS in the Mudumalai Tiger Reserve. Of which *Bambusa arundinacea* (grass species) was dominant for nesting of the IGS (11%, n = 22) followed by *Terminalia arjuna* (10%, n = 20), *Spondias mangifera* (9%, n = 18), *Syzygium cumini* (7%, n = 14). The previous study reported that in this region a total of 15 tree species were utilized for nesting purposes by IGS of which *Spondias mangifera* and *Schleichera oleosa* tree species were most preferable tree species for nesting [12]. The high preference for *Bambusa arundinacea and Terminalia arjuna*, *Spondias mangifera* and *Syzygium cumini* which are found mostly along rivers and streams could be due to their dense

	H	BA	DBH	BS	BE	BSBED	PB	SB	CL	CW	CC	DHH	DR	ELEV
H	1.000													
BA	−0.241	1.000												
DBH	0.014	0.901*	1.000											
BS	0.866*	−0.269	0.019	1.000										
BE	0.992*	−0.272	0.000	0.884*	1.000									
BSBED	0.876*	−0.177	0.019	0.546*	0.862*	1.000								
PB	0.012	0.692*	0.635*	−0.239	−0.047	0.207	1.000							
SB	−0.029	0.700*	0.593*	−0.044	−0.087	−0.050	0.570*	1.000						
CL	−0.053	0.785*	0.738*	−0.318	−0.126	0.151	0.886*	0.612*	1.000					
CW	0.862*	−0.123	0.025	0.501	0.825*	0.971*	0.292	0.056	0.256	1.000				
CC	−0.551**	0.292	0.117	−0.341	−0.518	−0.484**	−0.141	0.256	−0.119	−0.575**	1.000			
DHH	0.488*	0.245	0.376	0.441	0.469*	0.511*	0.136	0.405	0.187	0.476	0.691*	1.000		
DR	0.602*	0.369	0.436	0.481*	0.564*	0.596*	0.269	0.539*	0.360	0.623*	0.056	0.842*	1.000	
ELEV	0.224	−0.052	−0.054	0.154	0.176	0.200	−0.153	0.177	0.114	0.224	−0.114	0.216	0.254	1

Significant at $p < 0.05$.
**Positive correlation.*
***Negative correlation.*

Table 3.
Pearson's correlation coefficient matrix between nesting tree variables, by IGS in the Mudumalai Tiger Reserve.

Variables	Communality	PC1		PC2		PC3		PC4	
		r	c	r	c	r	c	r	c
Height	0.978	0.895*	0.153	−0.408	−0.094	−0.074	−0.042	0.076	0.069
BA	0.923	0.134	0.023	0.947*	0.217	−0.006	−0.004	0.087	0.079
GBH	0.794	0.323	0.055	0.798*	0.183	−0.060	−0.034	0.224	0.203
Branch Start	0.792	0.700*	0.120	−0.452	−0.104	0.235	0.134	0.208	0.188
Branch End	0.968	0.860*	0.147	−0.454	−0.104	−0.053	−0.030	0.142	0.129
Branch Start and Branch End Distance	0.884	0.852*	0.146	−0.285	−0.065	−0.270	−0.154	0.063	0.057
Primary Branch	0.916	0.292	0.050	0.722*	0.166	−0.553	−0.316	0.056	0.050
Secondary Branch	0.784	0.328	0.056	0.776*	0.178	0.242	0.138	−0.129	−0.117
Canopy Length	0.965	0.303	0.052	0.813*	0.187	−0.424	−0.242	−0.178	−0.162
Canopy Width	0.918	0.876*	0.150	−0.211	−0.048	−0.323	−0.184	−0.053	−0.048
Canopy Cover	0.815	−0.409	−0.070	0.373	0.085	0.639*	0.365	0.618*	0.288
Distance to Human habitation	0.801	0.714*	0.122	0.205	0.047	0.459*	0.256	0.219	0.198
Distance to Road	0.918	0.845*	0.144	0.287	0.066	0.331	0.189	0.106	0.096
Elevation	0.855	0.316	0.054	−0.039	−0.009	0.309	0.176	0.812*	0.736
Eigen value		5.85		4.36		1.75		1.10	
% Variance explained		39.02		29.07		11.68		7.35	
% Cumulative explained		39.02		68.10		79.77		87.12	

r- Pearson correlation coefficient, c- Factor Score coefficient
** Correlation significant at P < 0.05*

Table 4.
Summary statistics of principal component analysis.

S.No	Scientific name	Height (m)	DBH	Overall tree density/ha	Tree density/ha (GBH ≥ 250 cm)
1	*Acasia leucophloea*	8–26	96.57	1.2	0
2	*Alianthus excelsa*	5–28	93.44	8.91	0.26
3	*Albizia lebbeck*	6–	75.1	2.06	0
4	*Cassine glauca*	7–28	191.8	2.33	0.73
5	*Schleichera oleosa*	8–18	114.52	3.46	0
6	*Spondias mangifera*	8–25	228.61	2.46	0.93
7	*Syzygium cumini*	4–28	175.75	1.40	0.47
8	*Terminalia arjuna*	2–34	188.64	29.3	10.4
9	*Terminalia crenulata*	6–26	107.36	0.80	0.17
10	*Ficus benghalensis*	9–24	296.15	1.86	0.53
11	*Ficus microcarpa*	4–21	94.66	2.86	0.06
12	*Ficus mollis*	11–26	120.71	0.53	0.06
13	*Filicium decipiens*	8–20	134.17	1.13	0.13
14	*Manilkara hexandra*	7–28	150.73	4.40	0.6
15	*Sapindus emarginata*	8–16	96.5	0.86	0
16	*Tamarindus indica*	8–18	92.56	1.13	0.13
17	*Terminalia bellirica*	8–23	215.23	0.45	0.12
18	*Pongamia pinnata*	5–18	136.21	5.89	0.03
19	*Butea monosperma*	4–15	142.27	0.26	0
20	*Chloroxylon swietenia*	6–16	112.65	0.58	0
21	*Ficus racemosa*	10–23	286.12	0.34	0.05
22	*Ficus religiosa*	8–22	254.85	0.19	0.06
23	*Givotia rottleriformis*	5–13	116.57	0.68	0
24	*Holoptelea integrifolia*	6–12	154.36	0.75	0.02

Table 5.
Potential nest tree species, tree characteristics, and availability of IGS in MTR.

canopy cover, and higher canopy height and contiguity that could offer better protection and escape from predators.

A total of 192 nesting trees harboring 279 nests in an average of 2.66 nesting trees per km and 3.87 nests/km in a 72 km transect. In the previous study stated that a total of 83 nests were located along 54.2 km transects, giving an encounter rate of 1.5 nests/km of transects [12]. Previously the number of nests was reported in the moist deciduous forest [18] but in this study, I recorded the high number of nests in dry thorn forest riverine patches, it's evident that riverine patches afford good habitat for IGS in the environment. The assortment of nesting sites in most of the arboreal animal communities was seen in the riparian ecosystem, since of the diversity of plant species and tallness of the trees establish in these kinds of habitats and also accessibility of water for thermoregulation and humidity the stage of the enormous role for assortment of this habitat [19].

The nesting tree characters shows that the average height of the nesting tree and DBH and Trunk size and canopy had a very good percentage. Among the 19 nesting trees *Bambusa arundinacea* (Grass species), *Terminalia arjuna*, *Spondias mangifera* and *Syzygium cumini* trees contained the tallest height as well as DBH and Trunk size and canopy cover. The high preference for *Bambusa arundinacea*, *Terminalia arjuna*, *Spondias mangifera* and *Syzygium cumini* which are found mostly along rivers and streams, could be owing to their dense canopy cover, and higher canopy height and contiguity that could proffer better guard and escape from predators. Such prejudiced assortment towards matured trees with greater canopy contiguity could make easy group to and from the nest in all instructions, the main benefit to escape from predators and to move to other parts of the home range for foraging and other activities as reported by Ramachandran [8] Datta and Goyal [20] and Parathan [11].

The canopy length and width, as well as branch start and branch end, was very good in *Terminalia arjuna*, *Spondias mangifera* and *Syzygium cumini*, as well as these trees, hold most numbers of nest compare the other nesting trees. These results coupled with the results of nest tree characters show that the squirrels prefer the largest trees available and highest locations on the trees within their home range to build their nests. The selection is however strongly influenced by tree species and their physical characteristics including canopy contiguity as reported elsewhere [20] for the species. Prakash et al. [21] stated that the canopy length width is an important factor for choosing a nesting tree it provides shelter as well as protection.

This study found that a single tree holds a maximum five numbers of nest and minimum one nest and the average height of the nesting trees was 24.4 m. There were more than one or two nests in a single tree [18]. The tree species with multiple numbers of nests were *Terminalia arjuna*, *Spondias mangifera* and *Sizizyum cumini.* Kumara and Singh [22] sighted the Indian giant squirrel mostly at a height of 16 to 20 m in moist forests and 11 to 15 m in dry forests. We observed the Indian giant squirrel nesting on a large variety of the tree species (n = 37) in Karlapat wildlife sanctuary. Kanoje [23] also reported the use of a large variety of tree species (n = 30) for nesting in Sitanadi wildlife sanctuary, India. The nests were not built on the highest possible branch, as the squirrels sought cover above the nest. Such cover might help avoid direct heat from the sun and serve as hiding—place from birds of prey [10]. Among the 279 nests most of the nests were facing the northeast direction this is the influence of the sunlight effects plays a huge role in the nest position [24]. The maximum nests' width and length 70 and 35 cm, respectively. The nest condition, as well as length and width, play an important role in utilization as well as the care of young ones [7]. A nest was mostly located in the top (79%) and Middle (21%) of the canopy these results coupled with the results of nest tree characters show that the squirrels prefer the largest trees available and the highest locations on the trees within their home range to build their nests. The variety is however strongly predisposed by tree species and their physical characteristics as well as canopy contiguity as recorded in a different place [20] for the species.

6. Conclusion

Mudumalai Tiger Reserve faces severe pressure from the collection of non–timber forest products (NTFP) collection. Fruits of *Spondias mangifera* and *Tamarindus indica* are among the top NTFP collections and *Bambusa arundinacea* is highly utilized by local people for fences and home construction activities etc. which are also the preferred nesting trees of the Indian giant squirrel. The threats to this squirrel population in Mudumalai Tiger Reserve are immediate and visible. The results of this study support the need to implement the following conservation measures for the Indian giant squirrel: prevention of cutting *Bambusa arundinacea* of the preferred nesting species and regular monitoring of NTFPs; prevention of forest fires and mitigation of heavy grazing to allow regeneration of trees.

Mudumalai Tiger Reserve holds a good population of this endemic mammalian species in India. The sanctuary is a natural mosaic of different forest types and effective conservation management could have a positive, long-lasting impact on the population of the Indian giant squirrel.

Author details

Samson Arockianathan
Bombay Natural History Society, Mumbai, Maharashtra, India

*Address all correspondence to: kingvulture1786@gmail.com

References

[1] Corbet GB, Hill J. The Mammals of the Indo Malayan Region. Natural History Museum Publications, Oxford University Press Oxford, England, 1992. p. 488

[2] Molur S, Srinivasalu B, Srinivasalu C, Walker S, Nameer PO, Ravi KL. Status of South Asian Non–volant Small Mammals: Conservation Assessment and Management Plan (C. A. M. P.) Workshop Report. South Asia, Coimbatore, India: Zoo Outreach Organisation/ CBSG; 2005

[3] Molur S. *Ratufa indica*. The IUCN Red List of Threatened Species 2016: e.T19378A22262028. DOI: 10.2305/IUCN.UK.2016-2.RLTS. T19378A22262028.en [Accessed: 01 May 2020]

[4] Favre DS. International Trade in Endangered Species Guide to CITES. London: Martines Nijhoff Publishers; 1989

[5] Prater SH. The Book of Indian Animals. Mumbai: Bombay Natural History Society and Oxford University Press; 1971. p. 197

[6] Hayssen V. Patterns of body and tail length and body mass in sciuridae. Journal of Mammalogy. 2008;**89**(4): 852-873

[7] Borges R, Mali R, Somanathan H. The status, ecology and conservation of the Malabar Giant Squirrel *Ratufa indica*. Final report, Wildlife Institute of India. Indo-US Project; 1998

[8] Ramachandran KK. Certain aspects of ecology and behaviour of Malabar Giant Squirrel *Ratufa indica* (Schreber) [PhD Thesis]. Department of Zoology, University of Kerala; 1992

[9] Payne JB. Abundance of diurnal squirrels at the Kuala Lompat post of the Krau Game Reserve, Peninsular Malaysia. In: Marshall AG, editor. Abundance of Animals in Malaysian Rain Forests. United Kingdom: Department of Geography, University of Hull; 1979. pp. 37-51

[10] Pradhan AK, Shrotriya S, Rout SD. Observation on Nest- site selection by Indian giant squirrel in Karlapat Wildlife Sanctuary, Odish. Small Mammal. 2012;**4**(2):12-13

[11] Pradhan AK, Shrotriya S, Rout SD, Dash PK. Nesting and feeding habits of the Indian giant squirrel (*Ratufa indica*) in Karlapat wildlife sanctuary, India. Animal Biodiversity and Conservation. 2017;**40**(1):63-69

[12] Baskaran N, Venkatesan S, Mani J, Srivastava SK, Ajay DA. Some aspects of the ecology of Indian giant squirrel (*Ratufa indica* Erxleben, 1777) in the tropical forests of Mudumalai Wildlife Sanctuary, southern India and their conservation implications. Journal of Threatened Taxa. 2011;**3**(7): 1899-1908

[13] Champion HG, Seth SK. A Revised, Survey of the Forest Types of India. New Delhi: Govt. of India Press; 1968. p. 404

[14] James EC, Shugart HH Jr. A quantitative method of habitat description. Audubon Field Notes. 1970; **241**:727-736

[15] Kannan R. Ecology and conservation of the Great Pied Hornbill (*Buceros bicornis*) in the Western Ghats of Southern India [PhD Thesis]. USA: University of Arkansas; 1994

[16] Mudappa D, Kannan R. Nest site characteristics and nesting success of Malabar Grey hornbill in southern Western Ghats, India. Wilson Bulletien. 1997;**109**(1):102-111

[17] Girikaran P, Samson A, Ramakrishnan B, Ramasubramanian S. Nesting tree preference of Malabar pied hornbill (*Anthracoceros coronatus*) in pillur valley, Western Ghats, Southern India. Nature Conservation Research. 2019;**4**(3):45-53

[18] Srinivas V, Venugopal PD, Ram S. Site occupancy of the Indian giant squirrel *Ratufa indica* (Erxleben) in Kalakad Mundanthurai Tiger Reserve, Tamil Nadu, India. Current Science. 2008;**95**(7):889-894

[19] Archana M. Watershed Management. New Delhi: Author Press; 2001. p. 200

[20] Datta A, Goyal SP. Comparison of forest structure and use by the Indian giant squirrel (*Ratufa indica*) in two riverine forests of Central India. Biotropica. 1996;**28**(3):394-399

[21] Prakash S, Mishra AK, Raziuddin M. Studies on the nesting habits of Indian giant squirrel *Ratufa indica* centralis riley 1913 in Dalma Wildlife Sanctuary, Jharkhand, India. Journal of Life Science. 2011;**12**(12):9-18

[22] Kumara HN, Singh M. Distribution and relative abundance of giant squirrels and flying squirrels in Karnatak, India. Mammalia. 2009;**70**: 40-47

[23] Kanoje RS. Nesting sites of Indian giant squirrels in Sitanadi wildlife sanctuary, India. Current Science (Bangalore). 2008;7:882-884

[24] Nayak BK, Patra AK. Feeding and nesting ecology of Indian giant squirrel *Ratufa indica* (Erxleben, 1777) In Kuldiha wildlife sanctuary, Balasore, Odisha, India and its conservation. International Journal of Bioassays. 2015; **4**(03):3741-3746

Chapter 5

DNA Damage and Glutathione Peroxidase Activity in Liver and Kidney Cells in Wistar Rats Exposed to Terbuthylazine (TERB) for 28 Consecutive Days

Vilena Kašuba, Vedran Micek, Alica Pizent,
Blanka Tariba Lovaković, Davor Želježić,
Nevenka Kopjar and Mirta Milić

Abstract

The potential of low doses of the chloro-triazine herbicide terbuthylazine to induce DNA damage and impair activity of glutathione peroxidase (GPx) was evaluated in kidney and parenchymal and non-parenchymal liver cells of adult male rats. In a 28-day study, terbuthylazine was applied daily by oral gavage at doses: 0.004, 0.4 and 2.29 mg/kg bw/day. Tail Intensity (T Int) and Tail Length (TL) were used as descriptors of DNA damage. In the kidney, Tail Int was significantly different in all treated groups, while TL was different in 0.4 and 2.29 mg/kg bw/day groups, compared to controls. Significant differences in TL were recorded in parenchymal and non-parenchymal liver cells of all treated groups. Tail Int was significantly different from controls in non-parenchymal liver cells at all applied doses and in parenchymal cells at terbuthylazine doses of 0.004 and 2.29 mg/kg bw/day. A significant increase in GPx activity was observed only in the kidney at doses 0.4 and 2.29 mg/kg bw/day compared to the controls indicating its possible role in the protection of kidney from free radicals. It appears that repeated exposure to low doses of terbuthylazine could cause DNA instability in kidney cells and in parenchymal and non-parenchymal liver cells in rats.

1. Introduction

Terbuthylazine (TERB) is a chloro-s-triazine herbicide, mostly used for the removal of weeds to protect crops [1]. It is also used as an aquatic herbicide to control submerged and free-floating weeds and algae in fish ponds, swimming pools and reservoirs [2, 3]. It is the most frequently used triazine in Europe in the last two decades [4, 5]. The EFSA (2011) pointed out that TERB poses a high risk to non-target plants in the off-field areas, while the risk for soil micro- and macro-organisms and bees is low. Mammals could be exposed to TERB through oral,

IntechOpen

dermal or inhalation routes [6]. The acute toxicity of TERB can be low to moderate, causing slight eye and skin irritation and sensitisation. TERB shows adverse effects on the cellular activity of enzymes such as aromatase, an enzyme which converts androgen to oestrogen [6], and leads to cytotoxicity, as well as affects the functions of the kidney and liver [7, 8].

People are exposed to TERB in several ways, occupationally through inhalation and dermal contact at workplaces and at places where TERB is produced or used. The general population is mostly exposed through ingestion of contaminated drinking water or by dermal contacts [8].

This herbicide persists in the environment and easily moves from treated soil to water [9, 10]. The current cancer classification states TERB belongs to Group D "Not classifiable as a human carcinogen" [7].

Literature data shows that herbicides have the potential to induce reactive oxygen species (ROS), leading to oxidative stress on non-target organisms [11]. The first line of defence against the oxidative stress consists of the antioxidative enzymes superoxide dismutase (SOD), catalase (CAT) and glutathione peroxidase (GPx), which convert superoxide anions (O_2^-) into H_2O_2 and then into H_2O and O_2 [12–14]. Studies performed in experimental animals showed that atrazine and TERB mostly affected antioxidant defence of exposed animals, depending on species, its sex and age, herbicide concentration, and duration of exposure [15].

An US EPA study [8] indicated that TERB is highly toxic to aquatic organisms. It causes dose-dependent morphological changes and damage in gills, intestine, and kidney of fish [16], and disturbances in biochemical and oxidative stress parameters in carp [17]. Recent studies in aquatic organisms indicated that TERB can induce oxidative stress [18–20] and accumulation of ROS in cells [17, 21–23].

As for TERB genotoxicity, data are limited, i.e. no genotoxic effect was found using the bone-marrow micronucleus test both in female and male mice [24]. In an *in vitro* study, Mladinić et al. [25] showed that TERB could produce primary DNA damage in a 14-day extended human lymphocyte culture. Tariba Lovaković et al. [26] in *in vivo* study on rats exposed orally to TERB points to the disturbance of oxidant/antioxidant balance in erythrocytes and plasma. Plasma SOD activity decreased at two of the smallest doses (0.004 and 0.4 mg/kg bw/day), as well as plasma CAT activity at the highest applied dose (2.29 mg/kg bw/day).

The liver is the primary site of the metabolism, detoxification and excretion of potentially toxic substances [27]. The xenobiotic metabolism starts in the liver immediately after absorption from the gastrointestinal tract. One of the reasons for this is the fact that the liver has the highest supply of biotransformation enzymes of all organs in the body. Literature data [28] point to the fact that the liver is a complex organ with multiple cell types. Hepatocytes represent 60–65% of total rat liver cells [29]. They are responsible for the drug metabolism [30]. Their function is dependent on their micro environment, i.e., of direct cell-cell and cell-matrix interactions, and a lot of diffusible factors secreted by nearby non–parenchymal cells [31]. Non-parenchymal cells represent 40% of the total number of liver cells, but only 6.5% of its volume. These cells contribute to inflammatory responses. Several studies highlighted the importance of non-parenchymal liver cells and their responses to drug toxicity [32–36]. The main process of TERB metabolism in plants and animals is side-chain de-alkylation and oxidation to 2-hydroxy derivatives [37].

The aim of the present study was to investigate the impact of TERB on liver and kidney of adult male Wistar rats treated by oral gavage for 28 consecutive days. We determined levels of primary DNA damage by comet assay measuring Tail Length (TL) and Tail Intensity (T Int) in kidney and two types of liver cells: small or non-parenchymal cells (sized <30 μm head length) and medium sized cells or parenchymal cells or hepatocytes (sized between 30 and 40 μm head length).

We also measured the GPx activities in liver and kidney tissue to determine the extent of oxidative stress caused by the treatment. The results of the present study could contribute to understanding TERB's potential toxicity related to its low dose exposure.

2. Materials and methods

This experiment was conducted at the Institute for Medical Research and Occupational Health (IMROH), Zagreb, Croatia, in Animal Breeding Unit, Mutagenesis Unit and Analytical Toxicology and Mineral Metabolism Unit, in spring 2016.

2.1 Test substance and positive control substance

Terbuthylazine (CAS number 5915-41-3), purity grade 99.0%, was purchased as analytical standard from Pestanal® quality (Fluka, Sigma Aldrich, Germany) and dissolved in ethanol (EtOH) to prepare a stock solution. To prepare treatment solutions, the stock solution was diluted with sterile redistilled water. Ethyl methanesulfonate (EMS) was purchased from Sigma Aldrich, Germany.

Negative controls received water, whereas positive controls received ethyl methanesulfonate (EMS) at 300 mg/kg bw/day the last three days of the experiment. EMS is a monofunctional alkylating agent recommended for *in vivo* comet assay in rodents [38].

2.2 Animals

The study was approved by the Ethics Committee of the Institute for Medical Research and Occupational Health (IMROH), Zagreb, Croatia and the Croatian Ministry of Agriculture (Reg.no. 100-21/14-5, Class 01-18/14-02-2/6 of 11 June 2014). Animal treatments were carried out according to internationally accepted animal welfare guidelines [39]. The study was performed using 25 healthy male adult Wistar rats with an initial body weight from 231 g to 271 g. Free access to standard food (Mucedola, 4RF21, Italy) and tap water was ensured. Animals were kept in clear polycarbonate cages with 40–60% humidity at 22°C and normal 12-hr light/dark cycle. At the start of the study, the animals were weighted and inspected by a licenced veterinarian at IMROH.

2.3 Experimental schedule

Rats were randomly assigned to the five groups composed of five animals as recommended by JaCVAM (Japanese Center for the Validation of Alternative Methods) [40]. Three groups received TERB at doses of 0.004, 0.4 and 2.29 mg/kg bw/day, respectively, for 28 consecutive days by an oral gavage. These doses were selected based on the reference values set by the EFSA [6]. Negative controls received water, whereas positive controls received ethyl methanesulfonate (EMS) at 300 mg/kg bw/day over the last three days of the experiment. All animals were handled in the same manner.

Body weights were regularly monitored (once a week) during the experiment and the doses of TERB were adjusted accordingly. Survival and clinical signs of intoxication were also inspected daily by a licenced veterinarian at IMROH.

The experiment was terminated 24 h after the final gavage. All animals were humanely euthanized by exsanguination under Xylapan/Narketan anaesthesia (Xylapan, Vetoquinol UK Ltd., 12 mg/kg bw *i. p.*/Narketan, Vetoquinol UK Ltd., 80 mg/kg bw) and dissected. Animals were examined for gross pathological changes of the internal organs by a licenced veterinarian at IMROH.

At the end of each treatment, the body weight of rats was determined and compared with the initial body weight. Liver and kidney weight were also measured. Based on the obtained values, relative kidney (ROW_{kidney}) and liver weight (ROW_{liver}) were calculated using the following formula: ([organ weight/body weight at sacrifice day] × 100).

2.4 Slide preparation and the alkaline comet assay

Preparations of single cells were done within 1 h following sacrifice. Livers and kidneys were dissected and rinsed in cold TBS buffer (50 mM Tris-Cl, 150 mM NaCl, pH 7.5) [41] until as much blood as possible was removed. A small piece of tissue was put in chilly mincing buffer [75 mM NaCl (Kemika, Zagreb, Croatia) and 24 mM Na_2EDTA, pH 7.5] and minced with a pair of fine scissors to release single cells. The obtained cell suspension was stored on ice for a few seconds to allow large clumps to settle, and the supernatant was used to prepare agarose microgels for the alkaline comet assay. Slides were immersed in chilled lysing solution for at least 3 h in a refrigerator in the dark. Then the slides were rinsed in purified water to remove residual detergent and salts prior to the alkali unwinding step. Slides were randomly placed onto a platform of submarine type electrophoresis unit (Horizon 11.14, Whatman, Florham Park, NJ, USA) in the chilled electrophoresis solution, and left to unwind for 10 min. After 10 min of denaturation, the slides were electrophoresed at 1 V/cm for 10 min, with a constant voltage at approximately 300 mA, at +4°C [41]. The slides were then immersed in neutralisation buffer (0.4 M Tris buffer, pH 7.5) for at least 3 × 5 min. All slides were dehydrated by 70% ethanol and 96% ethanol (10 minutes each), air dried and stored at room temperature protected from humidity.

Slides were stained with ethidium bromide (20 μg/mL; Sigma, St. Louis, MO, USA) and analysed under an epifluorescence microscope (Olympus BX 50, Olympus, Tokyo, Japan), equipped with appropriate filters, under 200x magnification.

Three hundred cells (150 cells from each of two replicate slides, per each animal; five animals per group) were selected and analysed with a Comet Assay IV™ image analysis system (Instem-Perceptive instruments Ltd., Suffolk, Halstead, UK). All the comet measurements were performed on coded/blinded slides by the same person, experienced in scoring. When selecting cells, the areas around air bubbles or at the edges were avoided [42]. Two descriptors of primary DNA damage were selected, TL and a percentage of DNA in tail (T Int, expressed in % DNA).

Nucleoids with >80% DNA in the tail region were excluded from analysis of % tail DNA. They consisted of small or non-existent head and large, diffuse tails, and, according to literature data, they may represent DNA damage resulting from cytotoxicity [43, 44].

2.5 Determination of GPx activity in liver and kidney

Glutathione peroxidase (GPx) activity in liver and kidney supernatant was determined spectrophotometrically according to the European standardised method [45]. Briefly, 50 μl of liver/kidney supernatant was diluted with 500 μl of DL-Dithiothreitol (0.1 mol/L, Sigma, St. Louis, MO, USA). After 5 min of stabilisation, samples were further diluted (10 times) with double strength Drabkin's reagent and kept at 4°C until analysis. Portions of 0.8 mL of reaction mixture containing 0.1 mmol phosphate buffer pH 7.0, 0.01 mmol Na_2EDTA (Merck, Darmstadt, Germany), 1 EU glutathione reductase, 5 μmol of GSH and 0.25 μmol β-NADPH (Sigma, St. Louis, MO, USA) and 100 μl of diluted sample were pipetted into a measurement tube. The reaction was initiated with 0.1 mL t-butyl hydroperoxide (2.5 μmol, Sigma, St. Louis, MO, USA). The amount of GSH oxidised by t-butyl hydroperoxide was determined by following the decrease in the β-NADPH concentration, and the decrease in absorbance was measured at 340 nm (Cary 50 UV–Vis, Varian Inc. CA, USA). One unit of GPx is the number of micromoles of β -NADPH oxidised per minute. The results were expressed as IU/g protein.

Protein content was measured by Bradford assay [46] using bovine serum albumin (Sigma, St. Louis, MO, USA) as the standard.

2.6 Statistics

Statistical analysis was run using STATISTICA, version 13.2 (Dell Inc., Round Rock, TX, USA) software. Normality of data distribution was tested with Shapiro-Wilk's test. The data obtained for body and organ weight and GPx activity were normally distributed, while the data obtained for DNA damage were not. The results were expressed as means ± standard error, standard deviation, medians and ranges (min–max).

Data on weight of animals on the sacrifice day, liver and kidney weights, relative liver and kidney weights and GPx activity in both tissues were analysed with one-way ANOVA. For pairwise comparison, *post hoc* Tukey's HSD test was used.

Kruskal Wallis ANOVA by Ranks test with multiple comparisons (two tailed) was conducted to examine the differences in two descriptors of the alkaline comet assay (TL and T Int) and different herbicide doses. Statistical significance was set at $p < 0.05$.

3. Results

There was no incidence of mortality recorded in all the examined groups throughout the 28-day exposure period.

3.1 Body and organ mass, and relative organ weight

To assess the toxicity of applied TERB doses, the body mass of rats treated 28 days with 0.004, 0.4 and 2.29 mg/kg of TERB per day was compared with the mass of rats on the day 0 (the day before treatment). The results showed a significant reduction in a body mass of negative control compared to positive control (One-way ANOVA with Tukey's HSD *post hoc* analysis: $F = 5.94$, $df = 4$, $p < 0.001$) and 0.4 mg/kg bw/day group compared to positive control ($F = 5.94$, $df = 4$, $p < 0.05$) on sacrifice day. The weight gain of the rats exposed to TERB at all of the three applied doses was about 20% lower than the weight gain of negative control (**Figure 1**).

The effect of TERB on the organ weights and relative organ weights (ROWs) of the liver and kidney of adult male Wistar rats is shown in **Table 1**.

When compared liver mass, significant reduction in mass of PC compared to NC ($F = 5.89$, $df = 4$, $p < 0.05$), and 0.4 mg/kg bw/day group compared to PC ($F = 5.89$, $df = 4$, $p < 0.05$) was observed. The liver mass of the 0.004 mg/kg bw/day group was significantly lower than NC ($F = 5.89$, $df = 4$, $p < 0.05$) and the 0.4 mg/kg bw/day treated group ($F = 5.89$, $df = 4$, $p < 0.05$). When compared relative liver weights (ROW_{liver}), significant differences between the 0.004 mg/kg bw/day group with regard to PC, 0.4 mg/kg bw/day and 2.29 mg/kg bw/day group ($F = 4.53$, $df = 4$, $p < 0.05$) were observed.

There were no significant differences between the examined groups in kidney mass. A significant difference between PC and the 0.004 mg/kg bw/day group ($F = 4.62$, $df = 4$, $p < 0.001$) was found for relative kidney weight.

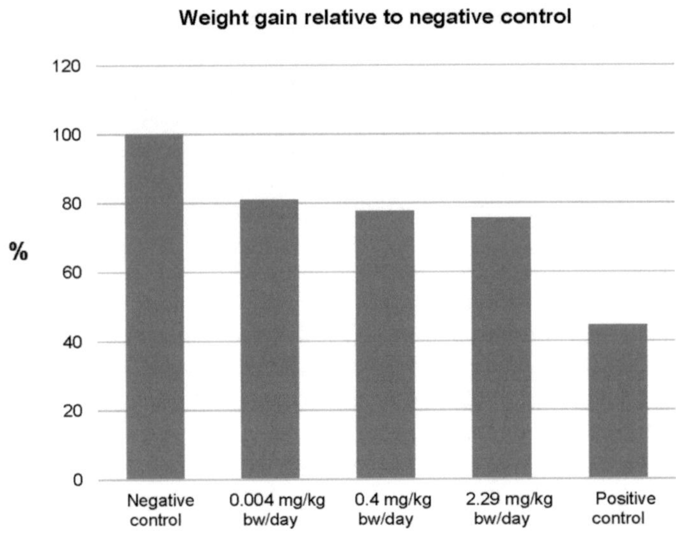

Figure 1.
Weight gain of the rats exposed to terbuthylazine at 28th day of consecutive exposure.

Organ	Negative control	Positive control[a]	Terbuthylazine 0.004 mg/kg bw/day	Terbuthylazine 0.4 mg/kg bw/day	Terbuthylazine 2.29 mg/kg bw/day
Animal weight on sacrifice day (g)	351 ± 10.89[PC] 24.35 366 312–368	289.2 ± 6.02 13.46 286 276–310	324 ± 5.61 12.55 318 314–345	333.6 ± 10.92[PC] 24.42 335 307–368	322.4 ± 11.06 24.72 327 286–352
Liver					
Organ weight (g)	10.35 ± 0,32[PC, b] 0.71 10.7 9.25–10.96	8.66 ± 0.37 0.82 8.57 7.58–9.76	8.79 ± 0.20 0.44 8.9 8.11–9.26	10.13 ± 0.51[PC,b] 1.33 10.48 8.96–11.6	9.77 ± 0.40 0.90 9.74 8.66–10.7
ROW_{liver}	2.95 ± 0.01 0.03 2.93 2.92–2.99	2.99 ± 0.09 0.20 2.91 2.75–3.24	2.72 ± 0.07[PC, c] 0.17 2.73 2.49–2.91	3.03 ± 0.06[b] 0.13 3.08 2.87–3.15	3.03 ± 0.05 0.10 3.02 2.92–3.2
Kidney					
Organ weight (g)	1.13 ± 0.04 0.09 1.15 1.01–1.24	1.01 ± 0.04 0.08 1.00 0.93–1.13	0.99 ± 0.03 0.07 0.99 0.93–1.11	1.07 ± 0.04 0.10 1.13 0.95–1.16	1.05 ± 0.03 0.08 1.03 0.96–1.15
ROW_{kidney}	0.32 ± 0.01 0.01 0.32 0.31–0.34	0.35 ± 0.01 0.01 0.35 0.33–0.36	0.31 ± 0.01[PC] 0.01 0.31 0.29–0.32	0.32 ± 0.01 0.01 0.31 0.31–0.34	0.33 ± 0.01 0.02 0.33 0.3–0.36

ROW_{kidney}: relative kidney and ROW_{liver}: relative liver weight.
The data are presented as mean ± SE (n = 5) and were evaluated by one-way ANOVA confirmed by Tukey's test. As an additional information, the data on standard deviation, median and range (min-max) were also obtained.[a]EMS(ethyl methanesulfonate) 300 mg/kg bw/day for the last three days of application.
[b]Statistically significant compared to 0.004 mg/kg bw/day.
[c]Statistically significant compared to 2.29 mg/kg bw/day.
[PC]Statistically significant compared to positive control.

Table 1.
The effect of terbuthylazine on the organ weights (mean ± SE) and relative organ weights (mean ± SE) (ROWs) of the liver and kidney of adult male Wistar rats.

3.2 Antioxidant response

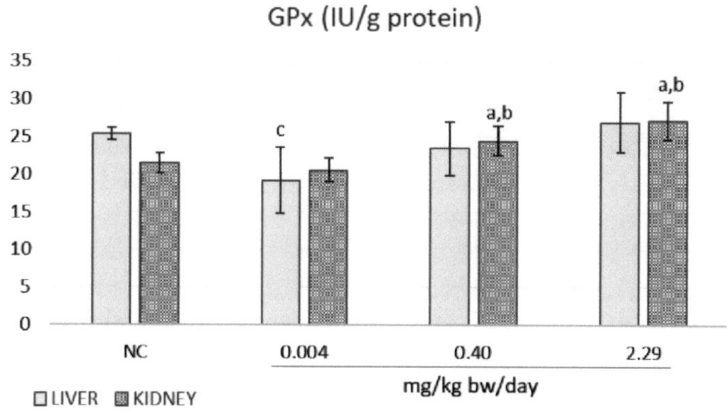

Figure 2.
The effects of the 28-day exposure to terbuthylazine on the activity of glutathione peroxidase (GPx) in the liver and kidney tissue of Wistar rats. Values are expressed as mean ± standard deviation of five rats per group (one way ANOVA followed by post hoc Tukey's HSD test for between groups comparisons, $p < 0.05$; a – significantly different from negative control (NC), b – significantly different from group treated with 0.004 mg/kg bw/day of terbuthylazine, c - significantly different from the group treated with 2.29 mg/kg bw/day of terbuthylazine).

	Kidney cells	
	Tail length ± SE SD Median Min.-max.	Tail intensity ± SE SD Median Min.-max.
Positive control EMS (300 mg/kg bw/day) for the last three days of treatment	29.97 ± 0.15^{NC,a,b,c} 5.33 30 12.92–54.17	14.51 ± 0.21^{NC,a,b,c} 7.28 13.77 0.01–39.29
Negative control	18.33 ± 0.11^{PC,a,c} 4.07 17.08 10–44.17	1.33 ± 0.08^{PC,a,b,c} 3.13 0.07 0–48.13
0.004 mg/kg bw/day	19.33 ± 0.16^{PC,c} 6.03 17.5 10.83–65.83	2.15 ± 0.11^{PC,NC,c} 4.31 0.13 0–39.74
0.4 mg/kg bw/day	19.74 ± 0.16^{PC,NC,c} 6.06 18.33 9.58–55.42	2.17 ± 0.11^{PC,NC,c} 4.32 0.17 0–38.71
2.29 mg/kg bw/day	20.78 ± 0.18^{PC,NC,a,b} 6.80 19.58 10.83–67.92	2.42 ± 0.11^{PC,NC,a} 4.17 0.22 0–26.63

The data were analysed using non-parametric Kruskal Wallis ANOVA by Ranks test with multiple comparisons (two tailed) for comparison between different treatment groups. Statistical significance was set at $p < 0.05$. ^{PC}*Significant compared to positive control.*
^{NC}*Significant compared to negative control.*
^{a}*Significant compared to 0.004 mg/kg bw/day.*
^{b}*Significant compared to 0.4 mg/kg bw/day.*
^{c}*Significant compared to 2.29 mg/kg bw/day.*

Table 2.
The comet assay parameters determined in kidney cells of rats exposed to terbuthylazine and corresponding controls.

In the current study, GPx activity in liver and kidney tissue homogenate was determined in order to assess the liver and kidney function and TERB-induced injury. The effect of 28 consecutive days of oral exposure to TERB on the activity of GPx in liver and kidney is presented in **Figure 2**.

A significant increase of GPx activity was observed in livers of animals treated with TERB ($F = 6.839$, $p < 0.05$, One way ANOVA followed by *post hoc* Tukey's HSD test for between groups comparisons). However, there was no significant difference between negative controls and treated animals. A comparison between groups showed a significantly higher activity of liver GPx in rats treated with 2.29 mg/kg bw/day of TERB compared to those treated with 0.004 mg/kg bw/day.

GPx activity significantly increased in kidney of animals treated with TERB ($F = 18.128$, $p < 0.001$, One way ANOVA followed by *post hoc* Tukey's HSD test for between groups comparisons). A comparison between groups showed a significantly higher activity of kidney GPx in rats treated with 0.4 and 2.29 mg/kg bw/day of TERB compared to negative control and rats treated with 0.004 mg/kg bw/day.

	Liver- non-parenchymal cells		**Liver-parenchymal cells**	
	Tail length ± SE SD Median Min.-max.	**Tail intensity ± SE SD Median Min.-max.**	**Tail length ± SE SD Median Min.-max.**	**Tail intensity ± SE SD Median Min.-max.**
Positive control EMS (300 mg/kg bw/day) for the last three days of treatment	$34.91 \pm 0.15^{NC,a,b,c}$ 5.83 34.58 10–61.25	$16.21 \pm 0.18^{NC,a,b,c}$ 6.97 15.72 0–37.60	$38.50 \pm 0.15^{NC,a,b,c}$ 5.86 38.33 14.58–77.5	$15.20 \pm 0.15^{NC,a,b,c}$ 5.92 15.07 0–38.48
Negative control	$18.40 \pm 0.13^{PC,a,b,c}$ 4.40 17.08 9.58–44.17	$1.19 \pm 0.08^{PC,a,b,c}$ 2.65 0.05 0–33.37	$22.54 \pm 0.14^{PC,a,b,c}$ 4.84 21.67 12.5–63.75	$1.53 \pm 0.09^{PC,a,c}$ 3.24 0.07 0–32.26
0.004 mg/kg bw/day	$20.21 \pm 0.13^{PC,NC,b,c}$ 5.06 20 7.08–47.08	$2.49 \pm 0.11^{PC,NC,b}$ 4.22 0.34 0–28.35	$23.79 \pm 0.14^{PC,NC,b,c}$ 5.32 23.33 11.67–52.5	$2.80 \pm 0.12^{PC,NC,b}$ 4.78 0.33 0–32.47
0.4 mg/kg bw/day	$21.79 \pm 0.16^{PC,NC,a}$ 5.52 20.83 11.25–59.58	$1.47 \pm 0.09^{PC,NC,a,c}$ 2.97 0.14 0–26.11	$25.39 \pm 0.13^{PC,NC,a}$ 4.79 25.42 10.83–54.17	$1.55 \pm 0.08^{PC,a,c}$ 2.99 0.09 0–18.49
2.29 mg/kg bw/day	$22.53 \pm 0.19^{PC,NC,a}$ 7.20 21.67 11.67–62.5	$2.31 \pm 0.11^{PC,NC,b}$ 4.07 0.41 0–32.39	$25.75 \pm 0.16^{PC,NC,a}$ 6.17 25.42 11.67–65.42	$2.51 \pm 0.11^{PC,NC,b}$ 4.29 0.35 0–34.10

The data were analysed using non-parametric Kruskal Wallis ANOVA by Ranks test with multiple comparisons (two tailed) for comparison between different treatment groups. Statistical significance was set at $p < 0.05$. [PC]Significant compared to positive control.
[NC]Significant compared to negative control.
[a]Significant compared to 0.004 mg/kg bw/day.
[b]Significant compared to 0.4 mg/kg bw/day.
[c]Significant compared to 2.29 mg/kg bw/day.

Table 3.
The comet assay parameters determined in parenchymal and non-parenchymal liver cells of rats exposed to terbuthylazine and corresponding controls.

3.3 The alkaline comet assay

Tables 2 and **3** report results regarding the comet assay parameters determined in kidney, and parenchymal and non-parenchymal liver cells of rats exposed to TERB and corresponding controls. Background levels of primary DNA damage in all tissues were low. Treatment with TERB at all three applied doses caused increased DNA instability in both tissues compared to negative control. The DNA of liver cells was more prone to breakage after TERB treatment compared to DNA in kidney cells. In liver cells, there was no dose-related increase of DNA damage, in contrast to kidney cells, where such a damage pattern was observed, both in terms of TL and T Int (**Table 2**). Inter-group differences in the levels of DNA damage measured in all cell types and their statistical significance are reported in **Tables 2** and **3**.

4. Discussion

This study aimed to assess the DNA damage and impairment of GPx activity in kidney and liver cells in a consecutive 28-day oral exposure of adult male Wistar rats to low doses of TERB. An EFSA document [6] pointed out that in mammals exposed through the gastrointestinal tract, animal skin, or inhalation routes, TERB acute toxicity can be low to moderate. Besides, it also causes slight eye and skin irritation, as well as sensitisation. Short-term exposure may affect body weight and food consumption in rats, mice, dogs, and rabbits. Long-term exposure may further affect organ weights in rats and mice, as well as haematological parameters in rats.

The results regarding body weight and organ weight changes observed after treatments with TERB in this study suggest that this herbicide at the tested doses and applied experimental schedule was able to produce acute toxicity, which led to changes in the overall fitness of the exposed *vs.* control rats. In toxicological studies, body weight and relative organ weight are widely accepted as a parameter associated with treatment-related effects. In official documents of regulatory agencies [6, 47] a significant decrease in body weight as the main effect of acute and long-term oral TERB exposure in experimental animal models was indicated. Such changes were usually connected with decreased food consumption. Since rats in this experiment had free access to food and water, treatment-related distress possibly reduced their appetite. Furthermore, the weight loss in exposed rats could also be related to its detrimental effect of treatment on intestinal absorption. Furthermore, the effects on liver weight reduction could be related both to functional liver deficiencies, but also to the impairment of different essential processes at cell level. It is possible that treatment produces loss of hepatocytes, due to cytotoxicity and apoptotic potential of the tested herbicide.

The environmental contaminants such as herbicides modulate antioxidant defensive systems causing oxidative stress, an abnormal phenomenon, which occurs in our cells or tissues when production of oxygen radicals exceeds their antioxidant capacity. Antioxidant enzymes catalyse the decomposition of ROS. GPx enzymes are the most important hydrogen peroxide (H_2O_2) removing enzymes in mammalian cells [48]. The liver is the major organ attacked by ROS [49]. Parenchymal cells or hepatocytes are primary cells subjected to oxidative stress-induced injury in the liver. To maintain the redox homeostasis in the liver, a sophisticated antioxidant system in mammals has been developed. Moreover, systemic oxidative stress arising during liver disease can also cause damage to the kidney [50]. Literature data suggests that systemic oxidative stress is considered to play a critical role in the pathophysiology of several kidney diseases [51, 52]. All parts of the kidney are affected by oxidative stress. Both directly and indirectly, vascular reactivity and renal hemodynamics, as well as glomerular filtration and tubular reabsorption and secretion in all nephron segments are included [53]. Disturbances in the antioxidant system could play a role in pathogenesis of chronic liver disease [54, 55]. GPx is one of the key enzymes in protecting the liver from the products of free radical reaction.

To the best of our knowledge, very little is known about the effects of TERB on oxidative stress parameters and antioxidant defence as well as primary DNA damage in mammals [15]. Studies on the influence of TERB exposure on parameters of oxidative stress are very rare. A few were performed in aquatic organisms such as the common carp (*Cyprinus carpio*) and red swamp crayfish (*Procambarus clarkii*). The results indicate that parameters of oxidative stress are altered under exposure to main degradation products of TERB at environmentally relevant concentrations, while TERB itself does not generally affect the oxidant/antioxidant balance in such conditions (reviewed in [15]).

In the current study, a slight disturbance in oxidant/antioxidant status was reflected in changes of the activities of GPx, mainly in the kidney. A significant increase in GPx activity was observed at 0.4 and 2.29 mg/kg bw/day compared to negative controls and animals exposed to 0.004 mg/kg bw/day of TERB. In the liver, no significant difference was observed between the negative control group and treated animals. However, significantly higher activity of liver GPx was observed in rats treated with 2.29 mg/kg bw/day of TERB compared to those treated with 0.004 mg/kg bw/day indicating that this effect could possess a certain toxicological risk at only slightly higher concentrations. The results suggest that repeated daily exposure to low doses of TERB stimulates the defending antioxidant mechanisms in order to alleviate the toxic effects of the produced reactive species.

It has previously been reported that triazine pesticides have a direct effect on kidney structure and function in freshwater fish [56–58]. The caudal kidney of common carps exposed to TERB showed alteration of tubular system of caudal kidney and the authors suggested that it was possible to describe TERB as a primary nephrotoxic substance. However, no similar study exists on rats with which we could compare our results.

In our earlier studies [15, 26], it was shown that TERB disturbs the oxidant/antioxidant balance in erythrocytes and plasma at the applied concentrations. Plasma SOD activity decreased at 0.004 and 0.4 mg/kg bw/day, as well as plasma CAT activity at 2.9 mg/kg bw/day. The observed increase of SOD activity in erythrocytes was most prominent at the highest applied concentration. An increase of whole blood GPx activity was observed at 0.4 mg/kg bw/day. Total antioxidant capacity expressed as plasma antioxidant power (FRAP) significantly increased at 0.004 and 0.4 mg/kg bw/day. In these experimental conditions, TERB did not induce lipid peroxidation.

The literature on primary DNA damage caused by TERB and related triazine herbicides in various cell types of *in vivo* exposed rodents is relatively rare. To the best of our knowledge, only a few studies used the alkaline comet assay as a method of choice in the evaluation of primary DNA damage in leukocytes of exposed mice or rats. Only a few studies have focused on *in vivo* [24] and *in vitro* experimental models [25, 59]. Using bone marrow micronucleus assay, Gebel et al. [24] did not find a genotoxic effect for TERB either in female or in male mice. In an environmental study in fish erythrocytes, Polard et al. [60] highlighted TERB as a potential water contaminant. Tennant et al. [61] tested the genotoxic effects of atrazine, simazine and cyanazine in mice following acute exposure at different doses, up to the maximum tolerated doses. They found relatively low genotoxicity in leukocytes. Singh et al. [62] studied the genotoxic effects of atrazine in male rats at the high dose of 300 mg/kg bw in 7, 14 and 21 days study. They reported a significant increase in comet TL in atrazine-exposed animals compared to controls, in blood and liver cells. In a study of prometryn on mice exposed to three doses for 28 days, Đikić et al. [63] found dose- and exposure-related DNA damage in the leukocytes.

As for DNA damage, our results of alkaline comet assay showed that the T Int in kidney cells was significantly different at all three doses as compared to controls, while TL was significantly different in 0.4 and 2.29 mg/kg bw/day groups compared to controls. In non-parenchymal liver cells, significant DNA damage (TL and T Int) was observed in all applied doses of TERB. Significant differences in TL were recorded in parenchymal liver cells at all applied TERB doses, while T Int was significantly different from controls at doses of 0.004 and 2.29 mg/kg bw/day.

The absence of a clear dose-response as seen from the T Int descriptor of the alkaline comet assay results (T Int in dose 0.004 mg/kg bw/day is greater than in 0.4 mg/kg bw/day dose, and in the highest applied dose) reported in **Table 2** may be explained by the fact that at higher TERB doses, actual DNA damage is possibly greater, and these highly damaged cells are lost from scoring. In this study, we excluded nucleoids whose TL exceeded 80, and head intensity was under 60. Thus, it is possible that the remaining cells that we measured by the image analysis system had less DNA damage. Such results could be the consequence of the presence of apoptotic cells. These cells have highly fragmented nucleoids which can be "washed" out from agarose gel during the comet assay processing. In such conditions, less damaged nucleoids will be measured, and the obtained values would be lower than real damage. The influence of apoptosis on DNA damage was established in Choucroun et al. and Roser et al. comet assay studies [64, 65].

The obtained results confirmed exposure-related genotoxicity in both types of liver cells. We could assume that a possible outcome of 28-day repeated exposure to TERB is a decrease in overall hepatic function.

Since there are no related comet assay studies on rats administered TERB in similar low doses, it is not possible to draw a parallel between our findings and other literature sources and propose the mechanisms behind the observed primary DNA damage.

In our earlier studies [26, 66], a slight induction of DNA damage in TL in rats leucocytes using alkaline comet assay, as well as reticulocyte frequency in MN *in vivo* assay, was observed. Two descriptors of the comet assay, TL and T Int were both lower at the two higher concentrations. Such results lead us to conclude that one of the possible mechanisms of action of TERB on DNA molecule could be intercalation, which could result in slower migration of DNA during electrophoresis. These effects could be explained by the ability of adaptation to the repeated doses of TERB. Kaware [67] suggested that liver tissue morphological changes during exposure to a toxicant could be adaptive mechanisms that allow animals to rapidly get rid of toxic compounds from the liver, through rapid metabolism and excretion, as sustained insults may lead to possible irreversible damages.

In a study on mice exposed intraperitoneally to TERB for 14 days at a daily dose of 0.0035 mg/kg, Želježić et al. [68] found a significant increase in mean TL and T Int in leucocytes, bone marrow cells, and liver cells compared to the control group. They found a significant increase of mean TL and T Int compared to controls in kidney cells of animals exposed to the formulated product Radazin TZ-50. Their results suggested that TERB metabolism possibly results in the formation of reactive metabolites capable of inducing DNA cross-links, which hinder DNA migration, and these effects were most pronounced in liver cells *in vivo*. The authors pointed to the fact that the differences in DNA damage between different cell types originated from the intrinsic metabolic differences between them.

5. Conclusions

From a toxicological point of view, this study shows that repeated *in vivo* exposure to low doses (0.004, 0.4 and 2.29 mg/kg bw/day) of TERB led to low-level DNA instability in kidney and non-parenchymal and parenchymal liver cells of adult male Wistar rats. An influence of applied low doses on GPx activity in kidneys was detected. Such results could point to the possible role of GPx as a key enzyme in kidney protection from hazardous products of free radical reactions. They can also reflect the response to increased oxidative stress.

Acknowledgements

This work was financially supported by Grant No. IP-2013-2111-8366 Organic Pollutants in Environment - Markers and Biomarkers of Toxicity (OPENTOX), funded by the Croatian Science Foundation.

Conflicts of interest

The authors declare that there are no conflicts of interest.

Author details

Vilena Kašuba[1*], Vedran Micek[2], Alica Pizent[3], Blanka Tariba Lovaković[3], Davor Želježić[1], Nevenka Kopjar[1] and Mirta Milić[1]

1 Mutagenesis Unit, Institute for Medical Research and Occupational Health, Zagreb, Croatia

2 Animal Breeding Unit, Institute for Medical Research and Occupational Health, Zagreb, Croatia

3 Analytical Toxicology and Mineral Metabolism Unit, Institute for Medical Research and Occupational Health, Zagreb, Croatia

*Address all correspondence to: vkasuba@imi.hr

References

[1] Wang H, Lin K, Hou Z, Richardson B, Gan J. Sorption of the herbicide terbuthylazine in two New Zealand forest soils amended with biosolids and biochars. Journal of Soil and Sediments 2010; 10: 283-289. doi: 10.1007/s11368-009-0111-z

[2] World Health Organization (WHO). Terbuthylazine in drinking-water. Background document for preparation of WHO Guidelines for drinking-water quality [displayed 4 June 2018]. Available at http://www.who.int/water_sanitation_health/dwq/chemicals/terbuthylazine.pdf

[3] World Health Organization (WHO). 2018. Terbuthylazine in drinking-water. Background document for preparation of WHO Guidelines for drinking-water quality. http://www.who.int/water_sanitation_health/dwq/chemicals/terbuthylazine.pdf

[4] Diez M.C. Biological aspects involved in the degradation of organic pollutants. J Soil Sci Plant Nutr. 2010. 10:244-267. DOI: 10.4067/S0718-95162010000100004

[5] Sass JB, Colangelo A. European Union Bans Atrazine, while the United States negotiates continued use. Int J Occup Environ Health 2006;12: 260-7. DOI: 10.1179/oeh.2006.12.3.260.

[6] EFSA - European Food Safety Authority (EFSA). Conclusion on the peer review of the pesticide risk assessment of the active substance terbuthylazine. EFSA J 2011; 9(1):1969. DOI:10.2903/j.efsa.2011.1969

[7] USEPA Office of Pesticide Programs. Chemicals Evaluated for Carcinogenic Potential. Annual Cancer Report 2017. http://npic.orst.edu/chemicals_evaluated.pdf

[8] US Environmental Protection Agency (US EPA). Reregistration Eligibility Decisions (REDs) Database on Terbuthylazine (5915-41-3). USEPA 738-R-95-005. 1995. https://archive.epa.gov/pesticides/reregistration/web/pdf/2645.pdf

[9] Calderon MJ, De Luna E, Gomez JA, Hermosin MC. Herbicide monitoring in soil, runoff waters and sediments in an olive orchard. Sci Total Environ. 2016;(569-570):416-22. DOI: 10.1016/j.scitotenv.2016.06.126.

[10] Stipičević S, Mendaš G, Dvoršćak M, Fingler S, Galzina N, Barić K. Dissipation dynamics of terbuthylazine in soil during the maize growing season. Arh Hig Rada Toksikol. 2017;68:336-42. DOI: 10.1515/aiht-2017-68-3063

[11] Ortiz-Ordonez E, Uria-Galicia E, Ruiz-Picos RA, Duran AGS, Trejo YH, Sedeno-Diaz JE, Lopez-Lopez E. Effect of Yerbimat herbicide on lipid peroxidation, catalase activity, and histological damage in gills and liver of the freshwater fish *Goodea atripinnis*. Arch Environ Contam Toxicol. 2011; 61:443-452. DOI: 10.1007/s00244-011-9648-0

[12] Ojha A, Yaduvanshi SK, Srivastava N. Effect of combined exposure of commonly used organophosphate pesticides on lipid peroxidation and antioxidant enzymes in rat tissues. Pesticide Biochemistry and Physiology 2011; 99(2):148-156. DOI:10.1016/j.pestbp.2010.11.011

[13] Seis H. Strategies of antioxidant defense. Eur J Biochem. 1993;215: 213-219. DOI:10.1111/j.1432-1033.1993.tb18025.x

[14] Tribble DL, Aw TY, Jones DP. The pathologic significance of lipid

peroxidation in oxidative cell injury. Hepatology 1987;7:377-387. DOI: 10.1002/hep.1840070227

[15] Živković Semren T, Žunec S, Pizent A. Oxidative stress in triazine pesticide toxicity: a review of the main biomarker findings. Arh Hig Rada Toksikol.2018;69:109-125

[16] Dezfuli BS, Simoni E, Giari L, Manera M. Effects of experimental terbuthylazine exposure on the cells of *Dicentrarchus labrax* (L.). Chemosphere 2006;64:1684-1694. DOI:10.1016/j.chemosphere. 2006.01.023

[17] Mikulíkova I, Modra H, Blahova J, Marsalek P, Groch L, Siroka Z, Kruzikova K, Jarkovsky J, Littnerova S, Svobodova Z. The effects of Click 500 SC (terbuthylazine) on common carp *Cyprinus carpio under* (sub)chronic conditions. Neuroendocrinol Lett. 2011;32(1): 15-24. PMID: **22167222**

[18] Stara A, Machova J, Velisek J. Effect of chronic exposure to prometryne on oxidative stress and antioxidant response in early life stages of common carp (*Cyprinus carpio* L.) *Neuroendocrinol Lett*. 2012;33(supplement 3):130-135. PMID: 23353856

[19] Stara A, Kristan J, Zuskova E, Velisek J. Effect of chronic exposure to prometryne on oxidative stress and antioxidant response in common carp (*Cyprinus carpio* L.) *Pest Biochem Physiol*. 2013;105(1):18-23. DOI:10.1016/j.pestbp.2012.11.002.

[20] Stara A, Machova J, Velisek J. Effect of chronic exposure to simazine on oxidative stress and antioxidant response in common carp (*Cyprinus carpio* L.). *Environ Toxicol Pharmacol*. 2012;33(2):334-343. DOI:10.1016/j. etap.2011.12.019

[21] Plhalova L, Stefanova S, Blahova J, Praskova E, Hostovsky M, Skoric M, Zelnickova L, Svobodova Z, Bedanova I. The effects of subchronic exposure to terbuthylazine on zebrafish. Neuroendocrinol Lett. 2012;33(3): 113-119. PMID:23353853

[22] Velisek J, Koutnik D, Zuskova E, Stara A. Effects of the terbuthylazine metabolite terbuthylazine-desethyl on common carp embryos and larvae. Sci Total Environ. 2016;539:214-220. DOI: 10.1016/j.scitotenv.2015.08.152

[23] Velisek J, Stara A, Zuskova E. Effect of single and combination of three triazine metabolites at environmental concentrations on early life stages of common carp (*Cyprinus carpio* L.). Environ Sci Pollut Res. 2016;23:24289-24297. DOI:10.1007/ s11356-016-7689-7

[24] Gebel T, Kevekordes S, Pav K, Edenharder R, Dunkelberg H. In vivo genotoxicity of selected herbicides in the mouse bone-marrow micronucleus test. Arch Toxicol. 1997;71:193-197. DOI:10.1007/s002040050375

[25] Mladinić M, Želježić D, Shaposhnikov SA, Collins AR. The use of FISH-comet to detect c-Myc and TP53 damage in extended-term lymphocyte cultures treated with terbuthylazine and carbofuran. Toxicol Lett 2012;211(1):62-69. DOI:10.1016/j. toxlet.2012.03.001

[26] Tariba Lovaković B, Pizent A, Kašuba V, Kopjar N, Micek V, Mendaš G, Dvoršćak M, Mikolić A, Milić M, Žunec S, Lucić Vrdoljak A, Želježić D. Effects of sub-chronic exposure to terbuthylazine on DNA damage, oxidative stress and parent compound/metabolite levels in adult male rats. Food Chem Toxicol. 2017;108:93-103. DOI: 10.1016/j. fct.2017.07.046

[27] Hall JE, Guyton AC. Guyton and Hall textbook of medical physiology. Philadelphia, PA; Saunders/Elsevier © 2011, 12th edition.

[28] Smith CC, Adkins DJ, Martin EA, O Donavan MR. Recommendations for design of the rat comet assay. Mutagenesis 2008;23(3):233-240. DOI:10.1093/mutage/gen008

[29] Stanger BZ. Cellular homeostasis and repair in the mammalian liver. Annu Rev Physiol. 2015;77:179-200. DOI:10.1146/annurev-physiol-021113-170255

[30] Malik R, Selden C, Hodgson H. The role of non-parenchymal cells in liver growth. Semin Cell Dev Biol. 2002;13:425-531. DOI:10.1016/s1084952102001301

[31] Kostadinova R, Boess F, Applegate D, Suter L, Weiser T, Singer T, Naughton B, Roth A. A long-term three dimensional liver co-culture system for improved prediction of clinically relevant drug induced hepatotoxicity. Toxicol Appl Pharmacol.2013;268:1-16. DOI:10.1016/j.taap.2013.01.012

[32] Jenne CN, Kubes, P. Immune surveillance by the liver. Nat Immunol. 2013;14:996-1006. DOI:10.1038/ni.2691

[33] McCuskey RS. Sinusoidal endothelial cells as an early target for hepatic toxicants. Clin Hemorheol Microcirc. 2006;34:5-10. PMID: 16543612

[34] Esch MB, Prot J-M, Wang YI, Miller P, Llamas-Vidales JR, Naughton BA, Applegate DR, Shuler ML. Multi-cellular 3D human primary liver cell culture elevates metabolic activity under fluidic flow. Lab on a Chip.2015;15:2269-2277. DOI:10.1039/c5lc00237k

[35] Bale SS, Golberg I, Jindal R, McCarty WJ, Luitje M, Hegde M, Bhushan A, Usta OB, Yarmush ML. Long-term co-culture strategies for primary hepatocytes and liver sinusoidal endothelial cells. Tissue engineering. Part C, Methods 2015;21:413-422. DOI:10.1089/ten.TEC.2014.0152

[36] Bale SS, Geerts S, Jindal R, Yarmush ML. Isolation and co-culture of rat parenchymal and non-parenchymal liver cells to evaluate cellular interactions and response. Sci Rep. 2016;6:25329. DOI:10.1038/srep25329

[37] Roberts TR, Hutson DH, Lee PW, Nicholls PH, Plimmer JR. Metabolic pathways of agrochemicals, In *Part 1:Herbicides and Plant Growth Regulators*, The Royal Society of Chemistry, Cambridge, UK, 1st edition, 1998.

[38] Pant K, Springer S, Bruce S, Lawer T, Hewitt N, Aardema MJ. Vehicle and positive control values from the in vivo rodent comet assay and biomonitoring studies using. Environ Mol Mutagen. 2014;55(8):633-642. DOI:10.1002/em.21881

[39] DIRECTIVE 2010/63/EU OF THE EUROPEAN PARLIAMENT AND OF THE COUNCIL of 22 September 2010 on the protection of animals used for scientific purposes. OJ L 276, 20.10.2010, p.33-79.

[40] JaCVAM (Japanese Center for the Validation of Alternative Methods), 2009. International validation of the *in vivo* rodent Alkaline Comet Assay for the detection of genotoxic carcinogens (version 14.2). Available at: http://cometassay.com/JaCVAM.pdf

[41] Peraica M, Ljubanović D, Želježić D, Domijan A-M. The effect of a single

dose of fumonisin B1 on rat kidney. Croat Chem Acta. 2008;81(1):119-124. ISSN-0011-1643

[42] Asahi J, Kamo H, Baba R, Doi Y, Yamashita A, Murakami D, Hanada A, Hirano T. Bisphenol A induces endoplasmic reticulum stress-associated apoptosis in mouse non-parenchymal hepatocytes. Life Sci. 2010;87:431-438. DOI:10.1016/J.LFS.2010.08.007

[43] McNamee JP, Bellier PV. Use of a standardized JaCVAM in vivo rat comet assay protocol to assess the genotoxicity of three coded test compounds; ampicillin trihydrate, 1,2-dimethylhydrazine dihydrochloride, and N-nitrosodimethylamine. Mutat Res Toxicol Environ Mutagen. 2015;786-788:158-164. DOI:10.1016/J.MRGENTOX.2015.02.005

[44] Lorenzo Y, Costa S, Collins AR, Azqueta A. The comet assay, DNA damage, DNA repair and cytotoxicity: hedgehogs are not always dead. Mutagenesis 2013;28:427-432. DOI:10.1093/mutage/get018

[45] Belsten JL, Wright AJ. European Community: FLAIR common assay for whole-blood glutathione peroxidase (GSH-Px); results of an inter-laboratory trial. Eur J Clin Nutr. 1995;49:921-927. PMID:**8925794**

[46] Bradford MM. A rapid and sensitive method for the quantitation of microgram quantities of protein utilizing the principle of protein-dye binding. Anal Biochem. 1976;72:248-254. DOI:10.1016/0003-2697(76)90527-3

[47] CLH, 2014. CLH Report Proposal for Harmonised Classification and Labelling Based on Regulation (EC) No 1272/2008 (CLP Regulation), Annex VI, Part 2. Substance Name: Terbuthylazine, EC Number: 227-637-9, CAS Number:5915-41-3.

[48] Halliwell B. Free radicals and antioxidants: A personal view. Nutr Rev. 1994;52:145

[49] Sanchez-Valle V, Chavez-Tapia NC, Uribe M, Mendez-Sanchez,N. Role of oxidative stress and molecular changes in liver fibrosis: A review. Curr Med Chem. 2012;19:4850-4860. DOI:10.2174/092986712803341520

[50] Palma HE, Wolkmer P, Gallio M, Correa MM, Schmatz R, Thome GR, Pereira LB, Castro VS, Pereira AB, Bueno A, de Oliveira LS, Rosolen D, Mann TR, de Cecco BS, Graca DL, Lopes STA, Mazzanti CMA. Oxidative stress parameters in blood, liver, and kidney of diabetic rats treated with curcumin and/or insulin. Mol Cell Biochem. 2014;386:199-210. DOI: 10.1007/s11010-013-1858-5

[51] Wang H, Chen X, Su Y, Paueksakon P, Hu W, Zhang MZ, Harris RC, Blackwell TS, Zent R, Pozzi A. p47phox contributes to albuminuria and kidney fibrosis in mice. Kidney Int. 2015;87:948-962. DOI:10.1038/ki.2014.386

[52] Valente MJ, Carvalho F, Bastos M, de Pinho PG, Carvalho M. Contribution of oxidative metabolism to cocaine-induced liver and kidney damage. Curr Med Chem. 2012;19:5601-5606. DOI:10.2174/092986712803988938

[53] Ratliff BB, Abdulmahdi W, Pawar R, Wolin MS. Oxidant Mechanisms in Renal Injury and Disease. Antioxid Redox Signal. 2016;25(3):119-146. DOI:10.1089/ars.2016.6665

[54] Czuczejko J, Zachara BA, Staubach-Topczewska E, Halota W, Kedziora J. Selenium, glutathione and glutathione peroxidases in blood of patients with chronic liver diseases. Acta Biochim Pol. 2003;50:1147-1154. PMID: 14740001

[55] Novo E, Marra F, Zamara E, Valfrè di Bonzo L, Caligiuri A, Cannito S, Antonaci C, Colombatto S, Pinzani M, Parola M. Dose dependent and divergent effects of superoxide anion on cell death, proliferation, and migration of activated human hepatic stellate cells. Gut. 2006;55:90-97. DOI: 10.1136/gut.2005.069633

[56] Velisek J, Stara A, Machova J, Dvorak P, Zuskova E, Svobodova Z. Simazin toxicity in environmental concentration on early life stages of common carp (*Cyprinus carpio* L.). Neuroendocrinol Lett. 2012;33(Suppl. 3): 90-95.

[57] Velisek J, Stara A, Koutnik D, Machova J. Effect of terbuthylazine-2-hydroxy at environmental concentrations on early life stages of common carp (*Cyprinus carpio* L.). Bio Med Res Int. 2014;(1):621304.

[58] Velisek J, Stara A, Koutnik D, Machova J. Effect of prometryne on early life stages of common carp (*Cyprinus carpio* L.).Pest Biochem Physiol. 2015;118:58-63.

[59] Mladinić M, Perković P, Želježić D. Characterization of chromatin instabilities induced by glyphosate, terbuthylazine and carbofuran using cytome FISH assay. Toxicol Lett. 2009;189:130-137. DOI:10.1016/j.toxlet.2009.05.012

[60] Polard T, Jean S, Gauthier L, Laplanche C, Merlina G, Sanchez-Perez JM, Pinelli E. Mutagenic impact on fish of runoff events in agricultural areas in south-west France. Aquat Toxicol. 2011;101(1):126-134. DOI:10.1016/j.aquatox.2010.09.014

[61] Tennant AH, Peng B, Kligerman AD. Genotoxicity studies of three triazine herbicides: in vivo studies using the alkaline single cell gel (SCG) assay. Mutat Res. 2001;493(1-2):1-10. DOI:10.1016/s1383-5718(01)00145-0

[62] Singh M, Kaur P, Sandhir R, Kiran R. Protective effects of vitamin E against atrazine-induced genotoxicity in rats. Mutat. Res. 2008;654(2):145-149. DOI:10.1016/j.mrgentox.2008.05.010

[63] Đikić D, Židovec-Lepej S, Remenar A, Horvat-Knežević A, Benković V, Lisičić D, Sajli L, Springer O. The effects of prometryne on subchronically treated mice evaluated by SCGE assay. Acta Biol Hung. 2009;60(1):35-43. DOI:10.1016/j.etap.2008.10.002

[64] Choucroun P, Gillet D, Dorange G, Sawicki B, Dewitte JD. Comet assay and early apoptosis. Mutat Res.2001;478:89-96. DOI:10.1016/s0027-5107(01)00123-3

[65] Roser S, Pool-Zobel B L, Rechkemmer G. Contribution of apoptosis to responses in the comet assay. Mutat Res.2001;497:169-175. DOI:10.1016/s1383-5718(01)00255-8

[66] Mikolić A, Žunec S, Kašuba V, Micek V, Milić M, Pizent A, Tariba Lovaković B, Lucić Vrdoljak A, Kopjar N, Želježić D. Terbuthylazine toxicity through a 28-day study: Oxidative stress responses and primary DNA damage in the blood of adult male Wistar rats. J Clin Toxicol. 2017;7:3(Suppl) DOI: 10.4172/2161-0495-C1-025

[67] Kaware M. Changes in liver and body weight of mice exposed to toxicant. Int Res J Sci Engin. 2013;3(1):92-95. ISSN:2322-0015

[68] Želježić D, Žunec S, Bjeliš M, Benković V, Mladinić M, Lovaković Tariba B, Pavičić I, Marjanović Čermak AM, Kašuba V, Milić M, Pizent A, Lucić Vrdoljak A, Kopjar N. Effects of the chloro-s-triazine herbicide terbuthylazine on DNA integrity in human and mouse cells. Environ Sci Pollut Res. 2018;**25**:19065-19081. DOI:10.1007/s11356-018-2046-7